T0329616

MORAL FIGURES

Making Reproduction Public in Vanuatu

In the early twentieth century, people in the southwestern Pacific nation of Vanuatu experienced rapid population decline, while in the early twenty-first century, they experienced rapid population growth. From colonial governance to postcolonial sovereignty, *Moral Figures* shows that despite attempts to govern population size and birth, reproduction in Vanuatu continues to exceed bureaucratic economization through Ni-Vanuatu insistence on Indigenous relationalities.

Through Alexandra Widmer's examination of how reproduction is made public, she demonstrates how population sciences have naturalized a focus on women's fertility and privileged issues of wage labour over women's land access, as well as broader social relations of reproduction. Widmer draws on oral histories with retired village midwives and massage healers on the changes to care for pregnancy and birth, as well as ethnographic research in a village outside the capital of Port Vila. Locating the Pacific Islands in global histories of demographic science and the medicalization of birth, the book presents archival material in a way that emphasizes bureaucratic practices in how colonial documents attempted to render Indigenous relationalities of reproduction governable.

While demographic imaginaries and biomedical practices increasingly frame fertility control as an investment in the reproductive health of individual bodies, the Ni-Vanuatu worlds presented in *Moral Figures* show that relationships between people, land, knowledge, kin, and care make reproduction a distributed and assisted process.

ALEXANDRA WIDMER is an associate professor of social anthropology at York University.

ANTHROPOLOGICAL HORIZONS

Editor: Michael Lambek, University of Toronto

This series, begun in 1991, focuses on theoretically informed ethnographic works addressing issues of mind and body, knowledge and power, equality and inequality, the individual and the collective. Interdisciplinary in its perspective, the series makes a unique contribution in several other academic disciplines: women's studies, history, philosophy, psychology, political science, and sociology.

For a list of the books published in this series see p. 245.

Moral Figures

Making Reproduction Public in Vanuatu

ALEXANDRA WIDMER

UNIVERSITY OF TORONTO PRESS
Toronto Buffalo London

Toronto Buffalo London
utorontopress.com

ISBN 978-1-4875-4320-4 (cloth)
ISBN 978-1-4875-4321-1 (paper)
ISBN 978-1-4875-4322-8 (EPUB)
ISBN 978-1-4875-4323-5 (PDF)

Library and Archives Canada Cataloguing in Publication

Title: Moral figures : making reproduction public in Vanuatu / Alexandra Widmer.
Names: Widmer, Alexandra, author.
Description: Includes bibliographical references and index.
Identifiers: Canadiana (print) 20220427518 | Canadiana (ebook) 20220427550 | ISBN 9781487543204 (cloth) | ISBN 9781487543211 (paper) | ISBN 9781487543235 (PDF) | ISBN 9781487543228 (EPUB)
Subjects: LCSH: Human reproduction – Social aspects – Vanuatu. | LCSH: Human reproduction – Economic aspects – Vanuatu. | LCSH: Human reproduction – Political aspects – Vanuatu. | LCSH: Reproductive health – Vanuatu. | LCSH: Vanuatu – Population policy. | LCSH: Vanuatu – Colonization – History. | LCSH: Childbirth – Social aspects – History. | LCSH: Medicalization – History. | LCSH: Demography – History.
Classification: LCC GN296.5.V36 W53 2023 | DDC 612.6099595 – dc23

We wish to acknowledge the land on which the University of Toronto Press operates. This land is the traditional territory of the Wendat, the Anishnaabeg, the Haudenosaunee, the Métis, and the Mississaugas of the Credit First Nation.

This book has been published with the help of a grant from the Federation for the Humanities and Social Sciences, through the Awards to Scholarly Publications Program, using funds provided by the Social Sciences and Humanities Research Council of Canada.

University of Toronto Press acknowledges the financial support of the Government of Canada, the Canada Council for the Arts, and the Ontario Arts Council, an agency of the Government of Ontario, for its publishing activities.

Funded by the Government of Canada | Financé par le gouvernement du Canada | Canada

Contents

Maps, Figures, and Tables

Maps

Figures

Tables

Preface

Do you remember how you felt in March and April 2020 – those months when COVID-19 reshaped social life the world over? Did you walk around your neighbourhood with an uneasy feeling in the pit of your stomach while avoiding people? Did you lose a dear friend or family member? Did you or someone you love lose their job? Did you wonder how you could endure the increased amount of work? Did you look around and ask yourself how things could continue like this? Did you constantly search online for more information about how bad things were, first in Wuhan then in Italy, Spain, Iran, and New York City, and compare the status of these places with that of your town or city? Did you long for times when you could get food and go to work without fear? Were you pregnant during this time? Or thinking about getting pregnant? Did you talk about what you would do when the pandemic was over? Did you think about family gatherings, restaurants, a regular workday? When did you start to long for things to go back to normal? Was it after a few days? A few weeks?

While COVID-19 has been experienced by some as unprecedented, pandemics are not new. This book begins at a moment in time in the early twentieth century after people in Vanuatu, an archipelago of over eighty islands in the south-west Pacific Ocean, had experienced several *decades* of deadly infectious disease epidemics. Some villages may have lost up to 90 per cent of their members. The archipelago's total population went from an estimated 650,000 (Speiser [1923] 1996, 39) or 500,000 (Spriggs 1997, 261) before sustained European contact in the 1850s to an estimated 65,000 people in the early twentieth century (Speiser [1923] 1996, 37). Demographers and other researchers have tended to gloss these decades of demoralization and fear as a period of "depopulation." It was a time when some Ni-Vanuatu[1] stopped using certain lands and deemed them *tabu*, and some converted or experimented with Christian values and social forms. Still others were kidnapped to work on plantations in Australia, while others sought out wage labour in Vanuatu.

This is a book about reproduction during and after decades of population decline. In my emphasis on the multiple dimensions, relations, and politics of reproduction after decades of infectious disease epidemics, I centre how people in Vanuatu persisted and engaged with new forms of knowledge, in this case biomedicine and demography, and the state forms of governance that rely on the quantification of life processes within populations. This broad conception of and emphasis on reproduction entails a recognition of the many relationalities that join people and groups together, an attention to how forms of labour are compensated, and an awareness of how land continues to matter as a foundation of life.

Acknowledgments

It is incredibly humbling to write these words of thanks. They form part of an ongoing recognition that any author works in an ecology of people who contribute in ways big and small. I am acutely aware that the acknowledgments are where a great deal of otherwise invisible labour and care of social reproduction are made public: thank you, one and all. The shortcomings, of course, remain mine.

My deepest thanks go to people in Vanuatu. I would like to thank Ralph Regenvanu, who was finishing his term as director of the Vanuatu Kaljoral Senta (VKS) at the time of my fieldwork, along with Henline Mala and Evelyne Buleigh at the VKS for their support. As well, I thank the members of the Vanuatu Cultural Research Council for approving my research permit. Profound thanks to Paramount Chief of Pango Rolland Maseiman Kalwatman, who granted permission for me to conduct ethnographic research in 2010. There are so many people who supported me and my little family in big and small ways during my fieldwork, and I am grateful and thankful to them all for sharing their experiences, knowledge, and hospitality. Some people whom I was fortunate to come to know in Vanuatu need special thanks: Edna Albert, Rossylin Isaiah, Leimara Jimmy, Augustine Kalmet, Ariel Kalotiti, Toupong Kalotiti, Lydia Kalran, Kirkir Kaltapang, David Kalsrap, Isobel Kalsrap, John Kalsrap, Julie Kalsrap, Kalfatak Kalsrap, Leinau Kalsrap, Martha Kalsrap, Monique Kalsrap, Sope Kalsrap, Edline Kaluatman, Martha Kaluatman, Elsie Lani, Alice Licht, Viviane Licht, Lucy Masing, Tousaro Samuel, Touro Samuel, Daina Simeon, Leiser Sogari, Vinau Sogari, Kalonuk Sope, Leisiel Sope, Naomi Sope, Tousaru Sope, Elinda Taleo, Leinawen Taleo, Nelly Taleo, Toumer Taleo, Toupet Taleo, Isaac Thomas, Louise Thomas, and Mercy Thomas. The people who were children in the much-beloved kindy, as well as the Kalsrap, Sope, and Thomas families, deserve many thanks for the hospitality they showed to Lucia and Rosalia van Schouwen.

I am grateful to Judith Littleton, Susanna Trnka, and especially Julie Park for their hospitality they showed me and my little family while in Auckland. This book owes so much to the incredible skill of the staff at the Western Pacific Archives at the University of Auckland. Thank you to Jo Birks, Nigel Bond, Ian Brailsford, Stephen Innes, and Katherine Pawley.

This book manuscript began to take shape during my postdoctoral position at the Max Planck Institute for the History of Science (MPIWG) in Berlin. This was an incredibly formative time for me. Veronika Lipphardt has my sincere thanks for her staunch support of me as an early career researcher with a young family and my respect for her wide-ranging intellectual curiosity and political sensibilities as regards history and science.

I am grateful to all of the scholars who spent longer and shorter periods of time in the Villa on Harnack Strasse at the MPIWG, and who contributed to a community concerned with the history of racism and human difference in the twentieth-century sciences of human variation. Samuel Coghe in particular understood the significance of demography in colonial endeavours, and I am grateful for our conversations, which were (and are) provocative and motivating. Thanks also to the other scholars whose engagement began during this time: Warwick Anderson, Barbara M. Cooper, Jenny Bangham, André Felipe Cândido da Silva, Joan Fuijimura, Ana Carolina Vimieiro Gomes, Geoffery Gray, Ricky Heinitz, Katrin Kleemann, Emma Kowal, Antje Kühnast, Ilana Lowy, Luci Luft, Staffan Müller-Wille, Amade M'Charek, Alondra Nelson, Letícia Galluzzi Nunes, Hans Pols, Joanna Radin, Katharina Schramm, Ricardo Ventura Santos, and Christine Winter.

Being a part of the Population Knowledge Network, expertly formed and led by Corinna Unger and Heinrich Hartmann, was incredibly helpful in terms of seeing how my work on demography in the Pacific connected to much larger historical questions and global issues. The intellectual gifts of the network were matched by gifts of conviviality at our meetings. Thank you to all the network members: Regula Argast, Samuel Coghe, Maria Dörnemann, Ursula Ferdinand, Heinrich Hartmann, Teresa Hühle, Axel Hüntelmann, Jesse Olsynko-Gryn, Petra Overath, Christiane Reinecke, Thomas Robertson, and Corinna Unger.

Over the long period of preparing this book, many conversations, scholarly and otherwise, have been my sustenance. Thank you to Susanne Bauer, Sarah Blacker, Margaret Critchlow, Maggie Cummings, Radhika Johari, Lynda Mannik, Jean Mitchell, Mary-Lee Mulholland, Karl Schmid, Martina Schlünder, and Justin Sully. Your friendship and engagement with ideas and drafts have provided the infrastructure for this book. It is impossible to think about having done this work without you.

More recent conversations about reproduction, medicine, land, and labour in Oceania have given me inspiration to complete the book. Thank you, Safua Akeli, Laura Burke, Leslie Butt, Heidi Colleran, Daniela Kraemer, Jacqueline Leckie, Madeleine Lemeki, Siobhan McDonnell, Jenny Munro, Victoria Stead, Chelsea Wentworth Fournier, Christine Winter, and Mingjen Wu.

I have been fortunate to have wonderful colleagues in the Anthropology Department at York University. I thank the members of the medical anthropology working group, especially Maggie MacDonald, for the generative conversations during isolating pandemic times.

Presenting parts of this work has brought welcome engagement. Thanks to audiences at meetings of the Association for Social Anthropology in Oceania, the European Society for Oceanists, the Pacific History Association, as well as the University of Amsterdam, the Centre for Ethnography at University of Toronto Scarborough, McMaster University, and York University. Special thanks to the organizers and discussants for their intellectual and logistical work, including Sandra Bamford, Maggie MacDonald, Amade M'Charek, Karen McGarry, Natasha Myers Jeannie Samuel, Katharina Schramm, Alice Servy, Holly Wardlow, and Donna Young.

Some material in chapter 1 was previously published in the 2014 article "The Imbalanced Sex-Ratio and the High Bride Price: Watermarks of Race in Demography and the Colonial Regulation of Reproduction" (*Science, Technology and Human Values* 39, no. 4 (July): 538–60, https://doi.org/10.1177/0162243914523509). Chapter 2 contains a handful of revised passages from the 2011 book chapter "Making Mothers: The Changing Relationships of Birth and Raising Children in Pango Village, Vanuatu" (in *An Anthropology of Mothering*, edited by Michelle Walks and Naomi MacPherson, 102–14 [Bradford, ON: Demeter Press]), as well as the 2014 book chapter "Filtering Demography and Biomedical Technologies: Melanesian Nurses and Global Population Concerns (1920–1970)" (in *A World of Populations: Transnational Perspectives on Demography in the Twentieth Century*, edited by Corinna R. Unger and Heinrich Hartmann, 222–42 [Oxford: Berghahn Books]). Earlier versions of portions of chapter 4 were previously published in two articles from 2013, "Of Temporal Politics and Demographic Anxieties: 'Young Mothers' in Demographic Predictions and Social Life in Vanuatu" (*Anthropologica* 55 [2]: 317–28) and "Diversity as Valued and Troubled: Social Identities and Demographic Categories in Understandings of Rapid Urban Growth in Vanuatu" (*Anthropology and Medicine* 20, no.

2 (August): 142–59, https://doi.org/10.1080/13648470.2013.805299). Thank you to the publishers for permission to reproduce these extracts here.

Many thanks to the undergraduate and graduate students at the Max Planck Institute for the History of Science and York University, who have been generous intellectual interlocutors and tenaciously kind with the details of this research and publishing journey. I wish to thank especially Massima Armani, Janelle Curry, Ricky Heinitz, Valaruthy Indran, Luci Luft, and Melina Schetakis.

At University of Toronto Press, Jodi Lewchuk has been a stellar editor to work with.

The anonymous manuscript reviewers have my everlasting gratitude for their incredibly astute, generous, and constructive comments. This was a gift beyond compare, made more precious for the fact that their time and knowledge were given during the first year of a global pandemic.

Thank you, as well, to friends and family who did not ask about the book during its long gestation, and especially Genevieve Curry, Ashley Fraser, Sarah Latha, Nancy Ng, Anna Purcell, Varya Rubin, Kiera Vanderlugt, Ernest Widmer, Evan Widmer, Karen Widmer, and Paul Widmer.

Jorge, Rosalia, and Lucia van Schouwen have made this book and our life surrounding it possible. You have taught me more than I could ever say about care and reproduction. Thank you.

MORAL FIGURES

Introduction

Daily routines of life begin early in Vanuatu. Roosters crow. Women light cooking fires or prepare milky sweet tea on a two-element propane cooking stove. By 7:00 a.m., if you live near Port Vila, the capital (population 50,944; VNSO 2016, 37), and you walk along the potholed main road from Pango Village (population 2,326; VNSO 2016, 37), you will see people dressed for work walking or waiting to flag down a red van that will take them the fifteen minutes' drive into town. They are going to their jobs at banks, grocery stores, development organizations, government offices, hotels, restaurants, or other people's homes. Older sisters brush their younger sisters' hair and prepare their uniforms for school. You might hear someone banging a bell for a church event.

Later on, if the tide is out, in Pango you might see women in their fifties and sixties expertly walking on the sharp coral reef with a slim metal rod looking for an octopus that might be hiding in a remarkably small hole. Men or women might be carrying food back from their gardens, which might be a few minutes' walk from their houses or reached via a longer, more rousing walk up the hill overlooking the bay upon which Port Vila is situated. This is the same hill from which American soldiers surveilled this bay during World War II, leaving enduring marks in the landscape. Women in Pango washed the American soldiers' clothes, a fact that is proudly memorialized in the Presbyterian church. Cultivated with people's care and skill, the gardens produce yams, manioc, kumala, pumpkin, and island cabbage, among other crops. Still, even if you have time to prepare the *aelan kakae* (island food) from the garden produce that you and your family grow, your children will probably prefer the white bread, rice, and lollies that can be bought at the store. Before and after school, and perhaps even during school hours, children chase dogs and run between houses in the village, while young people walk on the road in small groups, mobile phones in hand. Pango Village predates the colonial settlement of Port Vila. It has become a peri-urban place that prides itself on having been among the first to become Christian in Vanuatu. It ranks among those villages with the highest number of resorts in the country. There is an English school and a French school, small, family-run stores selling very basic household supplies and sweet or salty snacks, several churches, and a cemetery. Many houses have small gardens with manioc and island cabbage, and some others have a fruit tree or two. Garden lands with more substantial food production tend to be a ten- to thirty-minute walk away. Life, though far from easy, is vigorous. A very different picture of life in Vanuatu from my 2010 description was painted just over a century ago.

"Why should we go on having children? Since the white man came, they all die." The Swiss anthropologist Felix Speiser ([1923] 1996, 50) reported these words spoken by a Port Olry woman on the island of Espiritu Santo while conducting his fieldwork in Vanuatu from 1910 to 1912. In Speiser's estimation, women's inability to see a possible future in the face of social upheaval and widespread death had compelled them to use traditional means of controlling fertility. W.H.R. Rivers (1922, 104), the Cambridge psychiatrist and social anthropologist, recounted similar narratives: "The people say themselves: 'Why should we bring children into the world only to work for the white man?' Measures which, before the coming of the European, were used chiefly to prevent illegitimacy have become the instrument of racial suicide." In another example, a woman said to John Randall Baker, an Oxford biologist, "Close up all piccaninny here 'e die finish [*sic*]," or "nearly all the children here have died." Interpreting the women's sentiment as "apathy," Baker further opined that "this feeling that it is useless to produce children who will only die, is, I believe, the cause of small families on [Espiritu] Santo" (1928, 292). Researchers in the first decades of the twentieth century wrote about the social and biological phenomenon they called "depopulation" and debated whether it was due to "natural" causes (which were then couched in racial terms) or cultural or psychological ones. In the estimations of the anthropologists and biologist cited above, Ni-Vanuatu women's fertility control in a context of high mortality from infectious diseases demonstrated a lack of hope for the future. In publications and reports prompted by this rapid depopulation, these scholars generally understood the diseases brought by Europeans to be the cause of high mortality rates; low fertility rates were harder for them to explain, though researchers frequently blamed Ni-Vanuatu women's "insouciant" or "careless" attitude towards mothering (Jolly 1998, 183). The researchers were certain that these lives were tenuous and in need of care, or at least protection.[1] That the

Indigenous populations would die out was seen as a real possibility by colonial authorities, missionaries, and researchers.

By the 2010s, an entirely different demographic situation prevailed in Vanuatu. In Port Vila and Pango, the high number of young women giving birth, sometimes on mats on the floors of an overcrowded urban hospital, had become a public concern, voiced by demographers, medical staff, and the general public in Vanuatu. These young women who became mothers were sometimes chastised by members of the older generations because "they don't think about the future." This moral scolding is matched by an official discourse aimed at curbing pregnancy among young women. The force of this discourse was apparent during the 2010 Children's Day parade that I attended in Port Vila. The parade's theme of "Investing in the Nation's Future" emphasized education and biomedical health care as rights that were *stamba blong development long nesen ia, Vanuatu* (the foundation of the development of this nation, Vanuatu). Other banners contained phrases like *Mama and papa, no salem graon blong yumi!* (Mom and dad, don't sell our land!) and *Foreign education imas joinem witim kastom education blong gudfala fiuja blong nesen ia* (Foreign education must join with Indigenous education for a good future for this nation). Here, the moralizing disapproval some older Ni-Vanuatu can show towards young mothers about the future is recast in development discourse about education, amidst the competing discourse of preserving tradition and land. In 2014, the Vanuatu government conducted a survey (Vanuatu Ministry of Health, VNSO, and SPC 2014) that measured the "unmet need for contraception," a demographic measurement that tracks "the proportion of women of reproductive age (15–49 years) who are married or in a union and who have an unmet need for family planning" (WHO 2020). By 2017, demographers working at the United Nations Population Fund (which still goes by the acronym UNFPA, the United Nations Fund for Population Activities) and the United Nations Children's Fund (UNICEF) were writing about the need to realize the "demographic dividend" of a young and growing population in the Pacific Islands.

This is a book about social, economic, epistemic, and political relations of reproduction. I analyse these relations through episodes that unfolded over the century separating these two moments in Vanuatu's history, between what demographers and other researchers, governments, and international organizations have called depopulation in the early twentieth century and the rapid population growth of the early twenty-first century. Describing the shifting discourses and practices of reproduction over this period, I pursue the following questions: How is reproduction measured and quantified, and by whom? When is reproduction problematic? How are bodies governed, and by whom? Which reproductive behaviour and kinship practices are naturalized, and which are deemed changeable or contingent? Who provides the care? Who does the labour and who benefits? Whose knowledge matters? Answering these questions means showing that people can come to know, recognize, and act on the reproduction of individuals and populations through quantified representations of reproduction, and through biomedical and legal institutions that are founded with the intent of caring and governing. It means being attentive to the Indigenous socialities and forms of relationality that are entangled with colonial and postcolonial state forms of governance and, increasingly, with the knowledge production of international organizations. While "all politics are reproductive politics," as Briggs (2017) has convincingly argued of the US context, these are subject to certain local and historical inflections. Through five episodes in the history of Vanuatu, this book tracks the inflections of reproduction in quantification, medicalization, and care from their implications in colonial governance to postcolonial sovereignty.

When early twentieth-century researchers recorded Ni-Vanuatu women's despair about the drastic decline in the population, and when early twenty-first-century demographers enumerated population growth, their data – in narratives, anecdotes, quantifications – participated in a public envisioning of reproduction. Both the decline and growth of populations are demographic knowledge claims that shift life and death from the domain of individual or family concerns to that of the population. Throughout this book, I show how reproduction and social reproduction were made public in association with the circulation of quantified population knowledge and the practices of biomedicine. I focus on the political interpretations and mobilizations of biological reproduction and social reproduction through quantified population data and the associated medical and legal institutions. Although the practices of biomedicine, demography, and other quantified forms of population knowledge do not make all aspects of reproduction public, they are nevertheless increasingly significant in the period considered here in the forms that colonial and postcolonial politics take. In the first example I outlined above, which comes from the 1910s, the quantitatively measured population decline was interpreted politically according to a racial and colonial logic. In the second instance, from the 2010s, the population growth is in a sense leveraged, by the state, by NGOs, by international organizations, and by capital, in order to both discipline the reproductive practices of individuals and the population as whole, and neoliberalize subjects and social imaginaries.

A demographic enumeration of population dynamics is a precondition for both of these processes.

In the episodes I explore in this book, reproductive concerns, broadly construed, circulate publicly through the creation of what I call "moral figures." "Figures," here, denote quantitative knowledge of population that includes "imbalanced sex ratios," "subsistence," and "Melanesian well-being." At the same time, "figures" describe certain recognizable types of people, "nurse" or "young mother," whose social and symbolic importance emerges from their relationships to reproduction in Vanuatu. This dual meaning provides a term that can encompass both the enumeration, quantification, and representation of populations and their statistical behaviour for public circulation *and* the subjects that are constituted – as subjects, but also as actors or agents – out of efforts to operationalize this demographic knowledge and act on reproduction. "Young mothers" and "nurses" are thus both categories of people associated with the quantification and medicalization of reproduction and the subject and focus of a moral discourse. Through both kinds of figures, care, conduct, reciprocal relations, and distinctions between good and bad futures are negotiated in Vanuatu. They are thus moral in the way that Lambek (2015, ix) analyses ethics as everyday lived experiences of "reflection on how we live in relation to others (to past and present generations, to contemporaries and consociates, humans and non-humans)." The moral dimensions of the figures emerge in historically and socially particular ways in each chapter. Describing the construction and circulation of these moral figures is a way of showing different dimensions, relations, and politics of reproduction.

They are moral in at least two ways. First, the quantified moral figures circulate in association with the implicit or explicit categorization and governance of the right kinds of kinship, work, care, education, and economics in the social policies, institutions, and programs that are developed for state and capitalist forms of social planning. In the first part of the twentieth century, this connected reproduction with "civilization," and from the 1950s onward, with "economic development." As social identities, the figures are moral in that they associate development and change with gendered expectations of comportment and discipline. In a second sense, the figures are moral because they are associated with the quantification and subjectification of aspects of social life where relationality and reciprocal exchange are at stake. Relationality is a profoundly Pacific value emphasizing that right relations with others need cultivation and are at the moral centre of social life (e.g., Finau, Paea, and Reynolds 2022; Pala'amo 2019). In Vanuatu today, this is perhaps best encapsulated through the idiom of *respek* (e.g., Lindstrom 2017), a way of describing the right way of interacting with people, often mediated and made visible by exchanging objects, articulated in connection with Christianity and *kastom*. In this second dimension of "moral figures," my analysis thus highlights the Ni-Vanuatu relations of unpaid and paid care and work that are part of reproduction. Brought together in the way that I analyse both dimensions of "moral," I show how the relations of reproduction, moral in a Ni-Vanuatu sense, were also continuously quantified and governed through demography, biomedical institutions, and social policy inculcating different, but at times overlapping, moral relations. Through the circulations of these figures, I show that the values of colonialism and economic development were carried out through a moralizing of the relations of reproduction.

The processes of making reproduction public in Vanuatu had two important impacts. First, over the course of the twentieth century, the study and state governance of reproduction on the basis of demographic measurement increasingly located responsibility for population growth in women's bodies and individual behaviours. This discursive framing naturalizes women's role in reproduction and skews the representation of broader social structures and practices that shape reproduction, particularly the role of men and fathers, kinship systems, and access to land. The book will reveal the circumstances and measurements that have determined which gendered bodies, forms of care, and forms of kinship have been marked for improvement, and which ones have been stigmatized or overlooked in colonial and postcolonial contexts. Second, by examining how reproduction is made public, the book shows how reproduction is discursively and practically linked to multiple kinds of economic exchange. Thus, to examine how population thinking contributes to making reproduction public is to gain insights into colonial and postcolonial governance and the mechanisms of capitalist expansion. However, it is not only capitalist forms of exchange that are brought to bear in this process, for to examine how reproduction is made public in new ways also demonstrates the creativity of Ni-Vanuatu in forging Indigenous forms of modernity, including the maintenance of nonmonetary economies in the face of capitalist expansion.

A Brief History of Vanuatu

Ni-Vanuatu live on the over eighty islands that make up their south-western Pacific nation and speak over 110 languages. Before gaining independence in 1980, Vanuatu[2] (called the New Hebrides during the colonial period) was jointly administrated from 1906 by British and French authorities in an arrangement known as the Condominium. The islands were not a formal colony of either empire; instead, there were three streams of governance, British, French, and Condominium, as well as a "Native Code" developed for offences committed by Ni-Vanuatu that affected other Ni-Vanuatu. There was a British and French resident commissioner formally in charge in Port Vila with district agents stationed in the northern and southern regions (see maps 1–4).

Ni-Vanuatu (re-)achieved independence in 1980. The formation of anti-colonial political consciousness and movements began decades earlier and took place in several ways. There were Western-educated Ni-Vanuatu on the New Hebrides Advisory Council[3] who advocated for improved conditions from 1957 and whose activities accelerated in the 1960s. There were people participating in grass-roots anti-colonial movements led by Jimmy Stevens. There were Ni-Vanuatu clergy who, in meeting other Pacific Islanders during their training and at regional church conferences in the 1970s, learned of liberation-theological pedagogies as formulated by Ivan Illich and Paulo Freire and read anti-colonial writers like Franz Fanon (Gardner 2013). Such clergy were among the first generation of postcolonial politicians in Vanuatu and included the nation's first prime minister, Father Walter Lini, and the first cabinet minister of lands, Sethy Regenvanu. Britain and France had different approaches to formal legal decolonization; France was more invested in remaining and more reluctant to hand over control to Ni-Vanuatu leadership, while British authorities claim they had long been interested in withdrawing from the islands.

It has been estimated that prior to the population decline that followed the long-term settlement of Europeans in the mid-1800s,[4] the Indigenous population could have been between 500,000 (Spriggs 1997, 261) and 650,000 (Speiser [1923] 1996, 39). By the early twentieth century, the population is likely to have plummeted to around 65,000 (Speiser [1923] 1996, 37). Today, there is population growth,[5] accompanied by increased migration to urban centres. The population is now above 300,019 (VNSO 2021b), with rapidly increasing urban segments (see appendix 1 for more details of population size). Throughout the islands' unique colonial and demographic history, Ni-Vanuatu local practices and social life have undergone change while also being robustly maintained. Vanuatu is a place where over 80 per cent of the population do not participate in the formal economy (St-Hilaire of the Vanuatu Financial Centre Association, quoted in Bremner 2017). The obligations existing outside of the formal economy have both shifted and persisted through forms of colonialism, capitalism, and Christianity.

The social, political, spiritual, and material importance of land in Vanuatu cannot be overstated (e.g., McDonnell 2013, 2016; Rodman 1987;

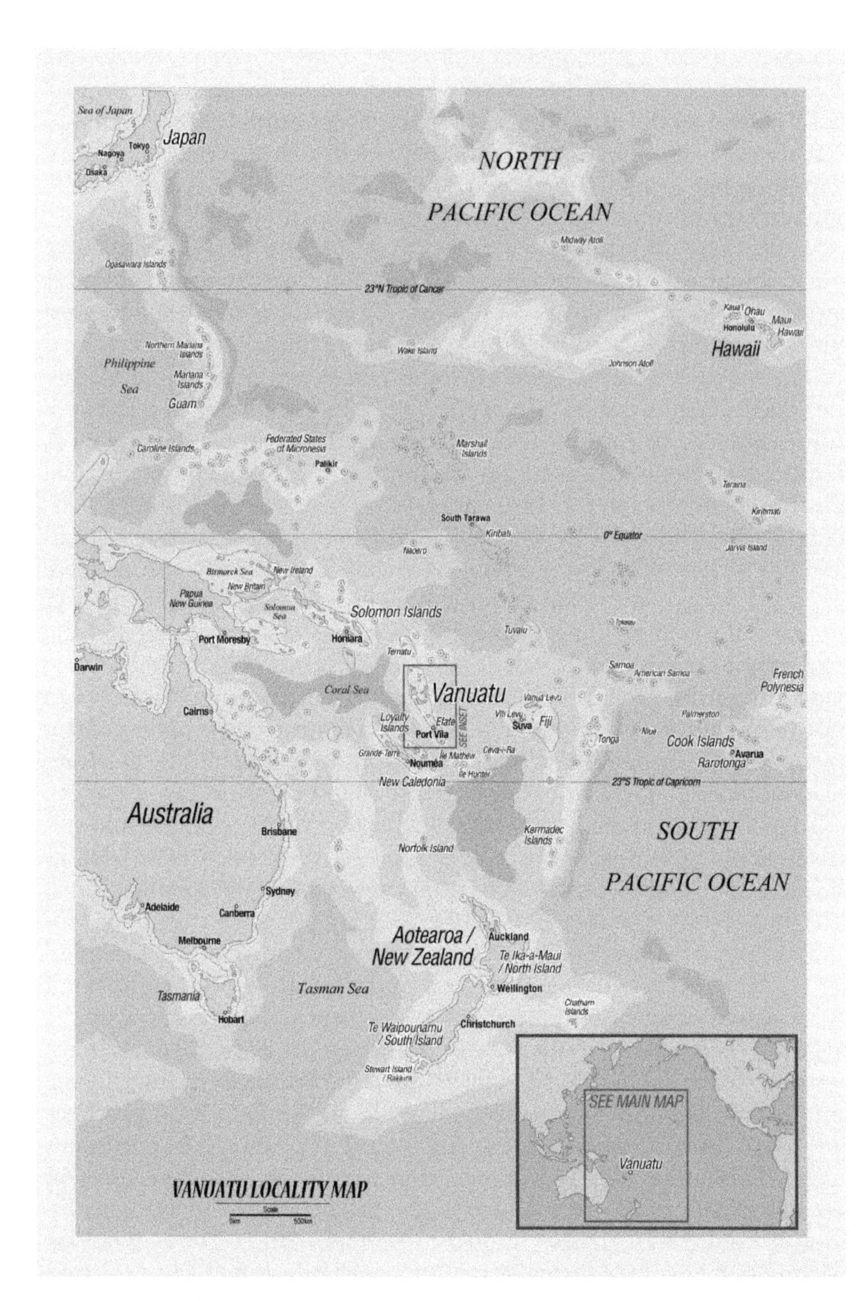

Map 1. Map of Pacific Ocean
Source: Map by Andrew Duggan with base map data © OpenStreetMap contributors. OpenStreetMap data is available under the Open Data Commons Open Database License.

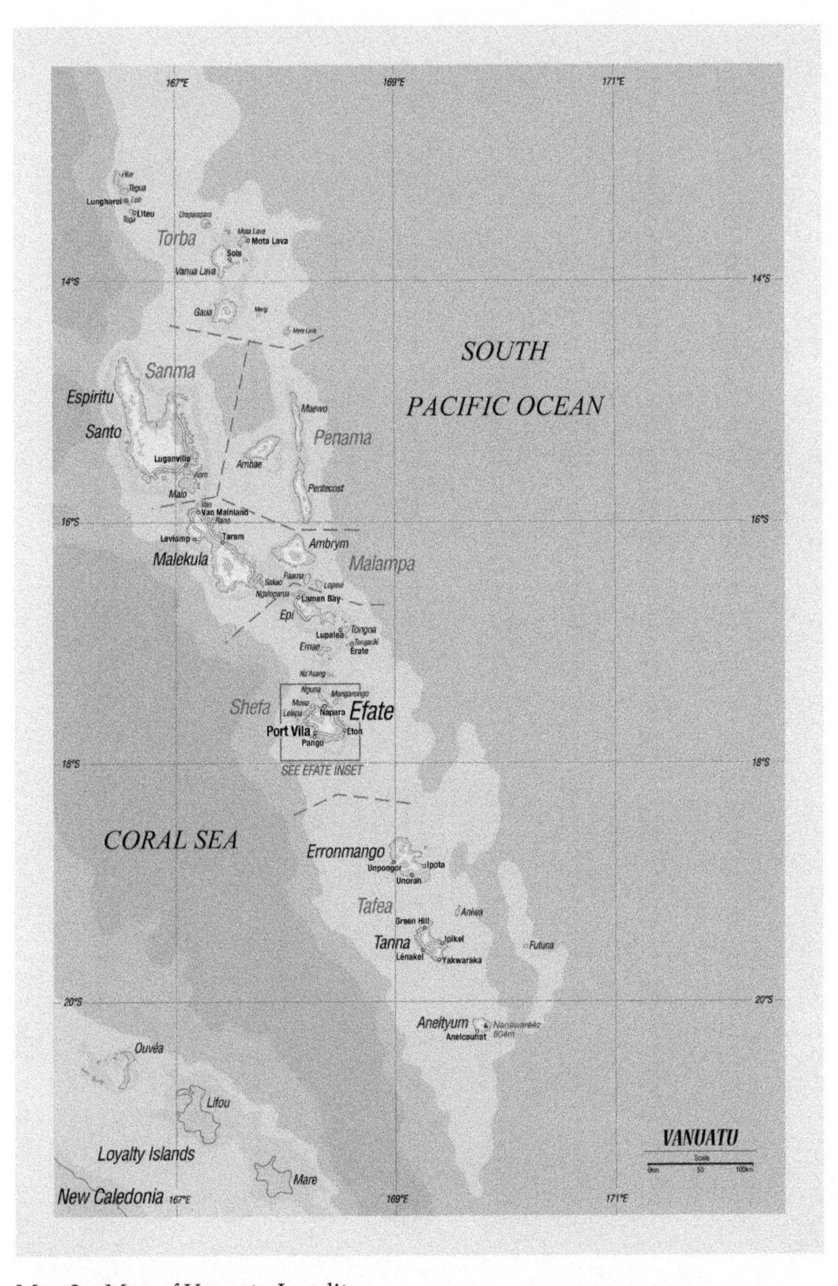

Map 2. Map of Vanuatu Locality
Source: Map by Andrew Duggan with base map data © OpenStreetMap contributors. OpenStreetMap data is available under the Open Data Commons Open Database License.

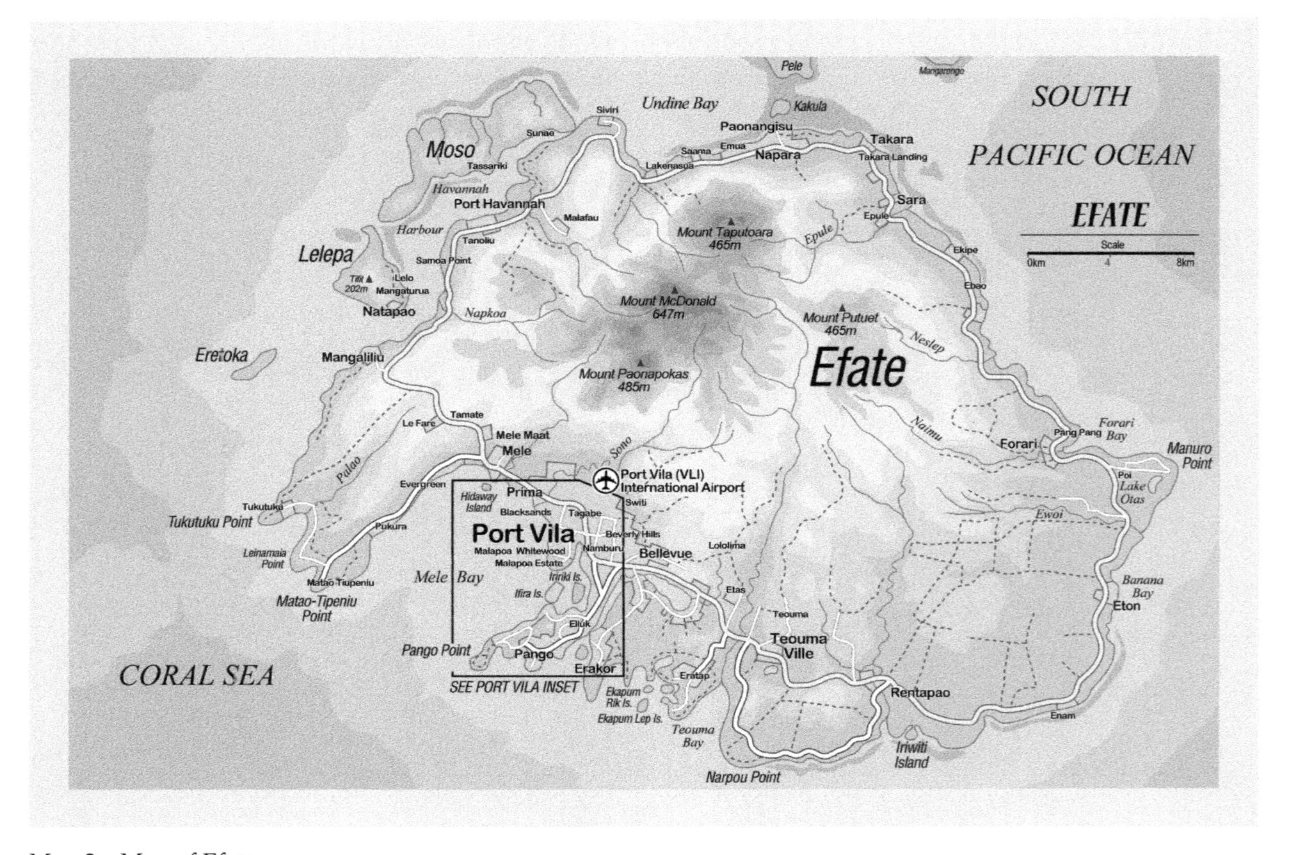

Map 3. Map of Efate

Source: Map by Andrew Duggan with base map data © OpenStreetMap contributors. OpenStreetMap data is available under the Open Data Commons Open Database License.

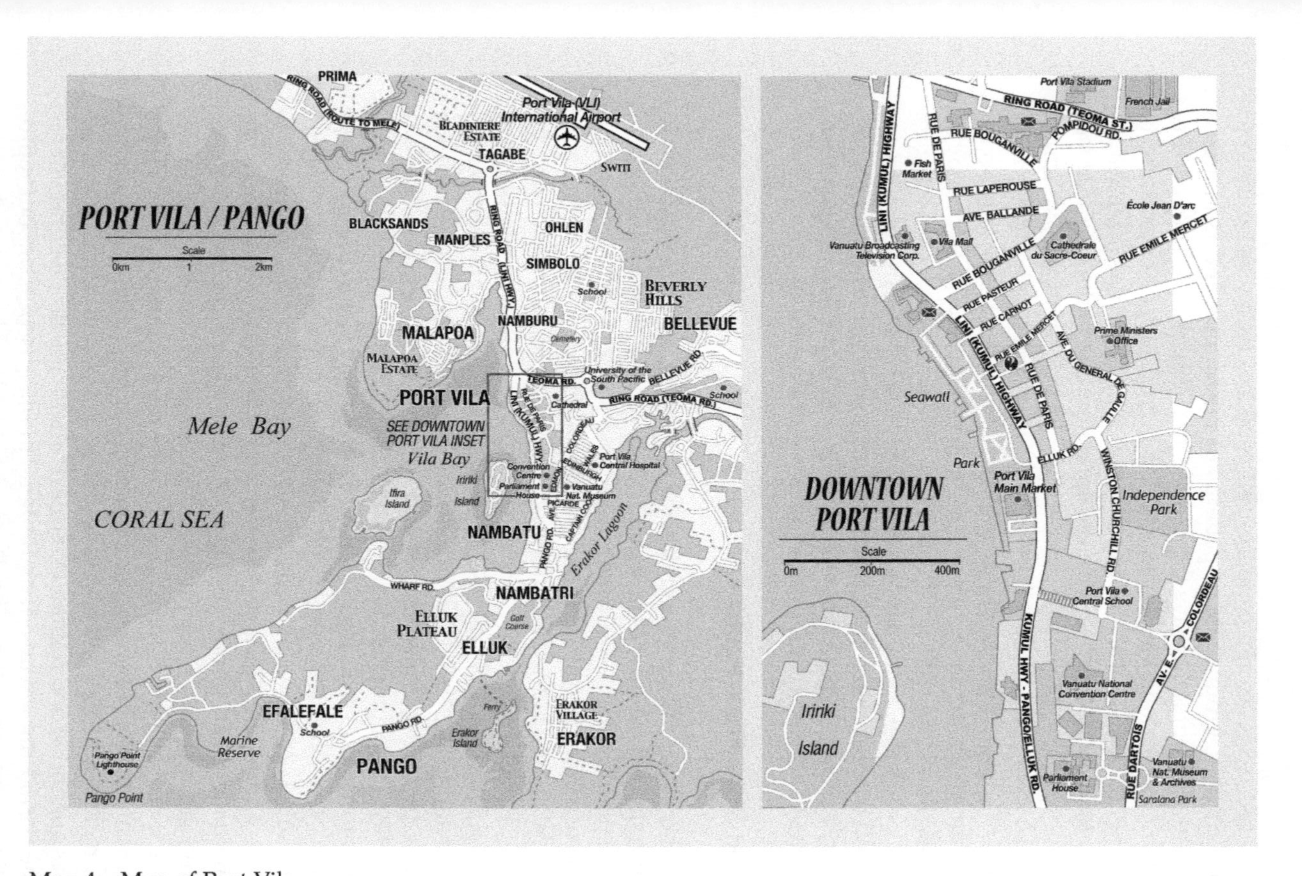

Map 4. Map of Port Vila
Source: Map by Andrew Duggan with base map data © OpenStreetMap contributors. OpenStreetMap data is available under the Open Data Commons Open Database License.

Van Trease 1987). At independence, all alienated land was returned to *kastom* owners, groups whose patterns of inheritance and land access are defined in different ways throughout the archipelago. Though a remarkable postcolonial achievement, this return of land, according to Rawlings (2002, 50), was not as universal as is sometimes claimed. In Pango, land inheritance and access are determined either by *blad laen* (bloodline) or *naflak* (usually glossed as "clan," an expansive group shaped by matrilineal connections). *Blad laen* refers to inheritance from fathers to sons, while inheritance through the *naflak* refers to land access based on group membership and passed generationally from mother to child. Rawlings (2002, 291), in his detailed analysis of land tenure systems in Pango, demonstrates that there has been considerable contestation over whether the *blad laen* or *naflak* should frame land ownership and access; some assert that the *naflak* is in decline, while others continue to advance claims based on these principles.

Reproduction Made Public

I use the term "public" throughout this book in an expansive way to name a space that emerges in Vanuatu in concert with quantified population knowledge, medical institutions, and forms of state governance. This space takes particular forms in association with, and is ever more entangled with, Indigenous ontologies and social infrastructures, including kinship networks, care economies, and other relationships at the local level. As such, my use of term is in line with anthropologists who examine the social, historical, and material processes by which publics are made through social and material practices as well as the circulation of documents (Anand 2018, 156). Making reproduction public in Vanuatu is associated with the creation of a demographic "population" that was also, under colonialism, an emergent polity subject to colonial laws – one that could be planned for in state governance. My use of the term "public" also names a formalized place where Ni-Vanuatu would encounter people from other villages, in the hospital or "native courts," and have problems addressed by outside authorities. It is a space subject to forms of expertise from European legal codes and biomedicine. There are also publics at various scales, including nation, region, and globe, where population is measured and reproduction is quantified and framed in terms of interventions. Over the course of the period under consideration here, public discourse about reproduction is increasingly medicalized, quantified, and connected to economic development. The colonial and then developmental discourse increasingly emphasizes individualized responsibility for one's life. By the 2010s, public responsibilities take the form of facilitating people's (especially women) ability to earn wages.

The connection between knowing about life processes within a population and governing the lives within that population is encapsulated in the conceptual terrain of "biopolitics." Foucault named this terrain as part of the "endeavor to rationalize the problems presented to governmental practice by the phenomena characteristic of a group of living human beings constituted as a population: health, sanitation, birthrate, longevity, race" (1997, 73). However, concerned as he was with liberalism, Foucault considered neither the colonial dimensions of governing populations, nor the formation of colonial subjectivities. Ann Stoler (1995, 2010) has pointed to the fact that the modern state and its documentary practices, especially censuses and statistics, which lie at the heart of Foucault's analysis of biopolitics and governmentality, do not readily translate to colonial or postcolonial contexts. This is particularly the case, Biruk (2018, 20) notes, with respect to racist assemblages in the measuring and management of would-be subject populations.

The way that states come to "see" people and social life (Scott 1998) has inspired a wide variety of scholarship on how people come to be seen through the deployment of various research methods (e.g., Law 2009). Analyses of the state have been nuanced, showing how its agents do not all "see" the state the same way (Blundo 2014). Furthermore, an understanding of how international organizations "see" (Broome and Seabrooke 2012) is also pertinent, especially when it comes to collecting survey data (Biruk 2012). What such works make clear is that the ability of states and international organizations to "see" depends on a diverse collection of people, infrastructures, and techniques. The desires and plans states have for ruling are not the same as how such rule is ultimately accomplished (Li 2007).

Critically attending to what is of public concern and what is of private concern has long been the practice of feminist anthropologists and anthropologists of reproduction and kinship. Collier, Rosaldo, and Yanagisako (1992) have shown that since the time of the European Industrial Revolution, families have been associated with the private sphere, which has been viewed in opposition to the public sphere in which individuals worked for wages, generally under exploitative conditions. The association of women with the family and domestic sphere and men with the economic sphere became central to European constructions of gender and gender inequality. These constructions constituted the basis for the claim made by anthropologists that the separation of domestic relations (associated with kinship or nature) and public relations (associated with politics, the economy, or culture) should not be presupposed; rather, they should be a matter of historical and critical ethnographic enquiry (McKinnon and Cannell 2013, 13). In various places, these anthropologists have shown that while modern institutions, such as those associated with capitalism or state governance, may purport to pertain exclusively to public and economic matters, rather than those considered private, they are in fact profoundly implicated in reproduction and kinship. Analysing how reproduction in this Pacific Island place is made public and institutionalized in colonial infrastructure reveals modern institutions' concern with kinship (e.g., McKinnon and Cannell 2013) in conjunction with quantifying life processes in a population.

Relatedly, anthropologists of reproduction have shown that reproductive concerns are not separate from public domains, but are instead rendered private or public in historically and socially specific ways. Though reproduction is often socially produced in relation to a historically specific private domain, many of the major public political battles of the twentieth century – such as conflicts over abortion, eugenics, and contraception – have been fought over reproductive concerns (Roberts 2015). This is a book about how reproduction is made public through quantification practices, medical institutions, and legal systems. Since quantifications of populations do not speak for themselves, highlighting the making of publics through moral figures also allows us to show that there is nothing "natural" or predetermined in how demographic figures are taken up.

Quantifying Reproduction: Demography and Population Thinking

The science of demography is fundamentally concerned with the statistical study of human populations, and especially their changing characteristics, behaviour, and structure over time. As such, demographers are interested in quantifying reproductive processes in populations. As well, because in demographic terms people enter a population through birth or migration, demographers are keenly interested in how people and populations move through space. Historically, the emergence of demography is connected to probabilistic thinking and the development of the European nation-state (Argast, Unger, and Widmer 2016) in the mid-nineteenth century, and by the mid-twentieth century it was associated with modernization and development (Doernemann and Huehle 2016). Population thinking quickly became something that could arouse concern in various contexts, from small villages, to colonial empires, to the entire planet (Bashford 2014), in concert with the development of technical infrastructures and organizations for counting individuals at different scales (Connelly 2008; Ferdinand and Overath 2016). Demographic knowledge informs policies on immigration and urbanization and influences geopolitics (Reinecke 2016). Demographers also contribute to knowledge and planning pertaining to environmental resources (Robertson 2016) and have influenced notions of the family and sexuality (Hartmann and Unger 2016). Demography thus has had a broad global influence by providing data and analysis for social policy and planning at multiple scales, but particularly in state governance.

Generating demographic knowledge is a social, political, and technical process. Relevant categories need to be decided on and target populations need to be defined. Members of these populations can sometimes decide whether to tell the truth to the enumerator or provide a fabricated answer, or they can use a variety of other methods to avoid answering entirely. Paper cards, tables, and calculation procedures need to be designed. This is part of a socio-technical process that Biruk (2018) refers to as "cooking data" in her ethnographic research with demographers in Malawi. While demography draws on disciplinary norms and assumptions, it depends on local infrastructures, socialities, and epistemologies.

Demography's frameworks and dominant concepts have changed over time (Population Knowledge Network 2016; Greenhalgh 1995, 1996). In the early twentieth century, racial conceptions of populations were prominent. The "uplift" or "degeneration" of a population through reproduction and the quantity and quality of populations were common eugenic concerns (e.g., Hartmann and Unger 2016; Solway 1990). The connection between racial thinking and demography was present in the first two decades of the twentieth century in the way that researchers in cosmopolitan research fora debated the imbalanced sex ratio in the Pacific Islands as a possible indicator of racial decline (Widmer 2012). As I show in chapter 1, this metric of the sex ratio, based on basic headcounts, circulated as a justification for how reproductive interventions should proceed in the early years of colonial rule. These included interventions in Indigenous cultural practices such as the "bride price." In the eyes of colonial administrators, the "bride price" should be lowered so that men could marry at a younger age.

In the mid-twentieth century, Frank Notestein and other American population experts revived and popularized the theory and associated categories of "demographic transition." Their theory divided the globe into populations that could transition from "traditionally" high fertility rates to low "modern" rates with appropriate economic modernization. Economic modernization meant wage labour. In chapter 3, I show

how the first census in Vanuatu was undertaken with an eye towards colonial planning within this economic-development logic. This was despite the fact that the census data demonstrated through the figure of "subsistence" that the vast majority of the population was already economically active, albeit not in a formal economy.

In the post-war era, global demographic discourses commonly claimed that a "population bomb" would result from a lack of appropriate economic development (McCann 2017). Campaigns aimed at population control (in the form of promoting biomedical contraception) were therefore launched to promote economic development and increase available land for food production (Robertson 2016). These demographic explanations connecting biomedical contraception, low birth rates, and modernity have been remarkably tenacious (Johnson-Hanks 2008), despite the ample contradictory evidence (Abernethy 1995). An important moment in the history of population thinking and policy was the UN-coordinated Fifth International Conference on Population and Development, held in Cairo in 1994. This event was evidence of a growing shift in population thinking from population control and economic development towards a focus on women's education and rights as a means of slowing population growth and facilitating economic development. This was not inconsistent with demographic transition theory, but, with an emphasis on women's education, this shift corresponded historically with a critique of the classical theory of transition that emphasized modernization. In the wake of this shift, the emphasis on educating and thus "investing" in girls grew as a dominant population paradigm within the NGO family-planning industry (Murphy 2017, 113). In current demographic thinking, as I discuss in chapter 4, "demographic dividends," "youth bulge," and the "unmet need for contraception" are common concepts and categories used to describe and analyse population growth in terms of investments and financial opportunity. New lives are deemed worthy of investment for their potential to generate individual financial returns.

There is a deeply problematic and enduring connection between racialization and population in demographic thinking and practices throughout the twentieth century, despite changes in focus and shifting scientific understandings of race (e.g., Murphy 2017; Widmer 2014). For example, a frequent theme within the circulation of demographic discourses was (and remains) the construction of racialized "others" as "too fertile" (e.g., Hartmann 1995). Very often the images and discourses of population growth "reinforce racial and ethnic stereotypes and scapegoat immigrants and other vulnerable communities" (Population and Development Program 2006). The association drawn between racialized population growth in the Global South and the associated use of environmental resources was entrenched in the intersecting discourses of demography and development. Most recently, population control and environmental concerns have once again been linked in population experts' discourse, this time in the connections made between population growth and climate change. Women's fertility and population growth in the Global South are blamed for the pressure they place on the environment, thereby contributing to climate change. The consumption patterns of people in the Global North are not subject to the same scrutiny in this discourse (Sasser 2018).

Population growth and decline are measured using demographic knowledge that is mobilized in the form of state intervention. Demography can support authoritarian state interventions aimed at reducing population size, as was the case, for example, in the "One Child Policy" (1979–2015) implemented in China (Greenhalgh 2008) and the often brutal expansion of contraception technologies during the "Emergency Period" (1975–77) in India (Olszynko-Gryn 2014; Tarlo 2003). Demographic knowledge has also guided state policies in Western European democracies whose aim, conversely, is increasing their populations (e.g., Krause 2005), or in the centrally controlled economies during the communist period in Eastern Europe (e.g., Kligman 1998). The history of Vanuatu shows how demographic thinking increasingly focused on women's agency in fertility control as a means of growing or reducing population size. The related colonial and postcolonial projects' reliance on women's agency was not matched by supportive resources; indeed the projects only worked through women's labour and Indigenous social networks.

Specific Quantifications: Census, Indicators, and Metrics

Demographic knowledge is collected and circulated in many ways; I focus on three in the following chapters. The modern census is a particular quantified rendering of a population that brings together the workings of the state and advances associated with probability thinking and statistical reasoning in the "avalanche of printed numbers" (Hacking 1993, 3). This work of categorizing and counting people to generate information about a population assumed particular salience in the nineteenth century. The census was closely tied to governmentality through nation states: the population categories delineated therein were groups of people whose welfare could ostensibly be improved through state intervention. In the early twentieth century, the emergence of new forms of state power over the lives and social welfare of a given population in Europe was also associated with civilizing agendas that justified the colonial presence. The census illustrates how "numbers gradually became more importantly part of the illusion of bureaucratic control and a key to a colonial *imaginaire* in which countable abstractions, both of people and of resources, at every imaginable level and for every conceivable purpose, created the sense of a controllable indigenous reality" (Appadurai 1997, 317). The desire for numbers did not always translate into the effective collection of census data by colonial state apparatuses, of course. Instead, medical doctors or scientists from many disciplines, including anthropology and biology, would often enumerate populations as part of their work. Missionaries would also collect census data and vital statistics in their church registries.

The role and aims of censuses in Vanuatu shifted over time. While British colonial administrators in early twentieth-century Vanuatu claimed that they did not have the resources to generate a census of Indigenous people (Widmer 2008, 2017), this had changed by the time of the post-war era. Ni-Vanuatu teachers, under the direction of the Australian demographer Norma McArthur and British administrator John Yaxley, conducted the first comprehensive census in 1967. In this census, the economic activities of "subsistence" – the unpaid, non-monetized labour of social reproduction in Vanuatu – dominated the accounts of economic activity. Knowing that most people did not work in the monetary economy, McArthur developed census questions to ascertain the non-monetized work of agricultural tasks and the communal exchanges of kin obligations. The resulting census showed that most people lived from the land and not wage labour, though wage labour participation was increasing. This was a census intended to plan for the population's health and education needs, and it further enrolled Ni-Vanuatu into the planning of colonial state politics. Related to these planning goals was the expansion of a labour force, a process that was seen as inevitable, despite the contrary evidence that "subsistence" sustained close to the entire population. This census siloed planning for wage labour from land use politics. Indeed, this was completely absent from the census (though it is present in other paper trails in the colonial archives), even though such concerns were ramping up at the time because of settlers' changes in land use for export production. Access to land, which most Ni-Vanuatu had at the time, was at the moral heart of Ni-Vanuatu worlds, and as such was vitally necessary for social reproduction, quantified as "subsistence."

Indicators are another means by which knowledge about populations is made available for particular ends. Merry argues that through "a wide range of quantitative and qualitative techniques for ordering knowledge" (2016, 12) indicators produce worlds where population information can be prioritized, ranked, and compared. Since indicators are constituted through techniques and categories of data collection, they do not reveal reality; rather, they produce it (Merry 2016, 32). Of note here is the contribution of indicators to governmental decisions because they render knowledge of social processes public in particular ways. For example, Merry (2016) writes of how the global problem of violence against women is translated into indicators of "severe" and "moderate" violence and thereby interpellated into targets for policing and global governance. The particular example of indicator culture that I examine in this book concerns the quantification of "well-being." The "alternative indicators" of well-being for Melanesia that I present in chapter 5, for example, quantify the importance of access to traditional building materials and traditional objects for exchange ceremonies, as well as opportunities to learn the languages of one's ancestors as a way of speaking back to GDP-oriented economic indicators of economic well-being.

Metrics are another form of quantification; they serve as "technologies of counting, but specifically technologies of counting that form global knowledge. Metrics are used today to offer uniform and standardized conversations about how best to intervene, how best to conceptualize health and disease, how best to both count and be accountable, and how best to pay for it all" (Adams 2016a, 6). Global health scholars (e.g., Adams 2016b) have made the important point that metrics shape what gets seen as a health issue. Global health practices mean collecting data and establishing "metrics to organize and make sense of that data, further surveillance and measures to determine whether interventions were successful and targets were met, and, increasingly, predictions that determine whether interventions will provide good returns on investments" (Brunson and Suh 2020, 1). Brunson and Suh show how, in practical processes of delivering maternal and reproductive health care, "metrics are imbued with meaning, morality, and power" (2020, 1). Metrics are thus political insofar as they shape what becomes the object of intervention and governance, and they are also moral because they are concerned with solving problems related to comportment, social values, and quantifying the consequences of right and wrong actions.

Population metrics and indicators have a place in colonial pasts and postcolonial presents. In Vanuatu, first colonial officials, followed later by Ni-Vanuatu government representatives, mobilized metrics and indicators and their quantified framings of (social) reproductive processes to explain social problems, justify interventions, evaluate progress on initiatives, or assert policy alternatives. The entanglement of metrics and indicators and the assertion of authority in the changing social forms of state governance is apparent. "Imbalanced sex ratio" and "well-being for Melanesia" are, respectively, a metric and an indicator adopted at different historical moments in Vanuatu. The imbalanced sex ratio – there were far more men than women – measured the population and provided a crude metric (though that term was not used) that indicated population health in the early twentieth century. As I show in chapter 1, the problem of population decline indexed by this metric mandated colonial attempts to intervene on marriage practices in an effort to enable more men to marry at a younger age. "Well-being for Melanesia" is an alternative indicator developed in postcolonial Vanuatu that quantified and rendered social reproduction visible in terms of the aspects of culture mainly associated with traditional economies that centre Ni-Vanuatu values of reciprocity. This quantification, which makes social reproduction public, serves not only as an external technology of governance, but also as a means of asserting sovereignty through an explicit critique of the kinds of economies associated with contemporary development. This indicator is an Indigenous critique of overly monetized relationships and their impacts on well-being, just as it challenges our thinking about the statistical measurement of well-being. Yet, as I will show, since they are explicitly designed to speak to conventional economic indicators, alternative indicators do elide certain kinds of knowledge and care associated with well-being.

In outlining the frameworks of census, metrics, and indictors, I hope to make clear – to borrow the title of Lisa Gitelman's (2013) book – that *"Raw Data" Is an Oxy-Moron*. In other words, population data is always shaped by the way it is gathered and represented. The census, metrics, and indicators analysed in the following chapters do not only quantify reproductive processes, they also frame the ensuing quantitative representation and suggest narratives in the public circulation of those figures. Quantifications are associated with narratives and explanatory solutions as they are rendered public knowledge. It is the content, representations, and the public circulations of these quantified figures and their entanglements with Ni-Vanuatu worlds that I present this book; I do not systematically explore the validating interactions of public expert cultures, scientific methods, and quantification in the making and legitimation of the metrics and indicators.[6] Through the circulation of figures, reproductive processes become public knowledge; metrics and indicators enable and participate in public discourses of reproduction that are ever more economistic and monetized. Yet it is also the case that, with the ingenuity of individual Ni-Vanuatu and robust social networks, these figures offer spaces in which to apprehend and assert sovereignty.

Reproductive Institutions: Courts and Hospitals

Demographic research is associated with institutions that medicalize reproduction, bringing its biological aspects under the purview of doctors, nurses, and hospitals. Demographic figures also emerge out of institutional contexts that privilege a medicalized understanding of reproduction, likewise framing it as a primarily biological process under the purview of doctors, nurses and hospitals. Public or mission hospitals in European colonies have been sites that legitimate and enable colonial power, in addition to promoting modernization and developmental agendas. Bringing birth and infant care into their ambit has been an integral aspect of hospital work. The incorporation of birthing women and women's reproductive bodies into colonial agendas was implicated in a much broader process with the attempt to inculcate Western notions of the body, motherhood, and sexuality within subject populations (e.g., Boddy 2007; Ram and Jolly 1998; Guha 2018; Coghe 2022; Hunt 1999). In the 1950s and 1960s, Presbyterian and Anglican missionaries formalized the nurse training program for Ni-Vanuatu women with the support of the British colonial government. The nurses trained through these programs – among the first Ni-Vanuatu women professionally educated in a Western educational system – acted as intermediaries between Indigenous and biomedical knowledge paradigms and forms of social organization relating to birth, fertility, and health. Their oral histories show care as embedded in relational infrastructures. When in the care of these nurses, pregnant Ni-Vanuatu women in the area around Port Vila did not see biomedicine as an entirely intrusive measure for birth in the 1950s and 1960s; they were receptive to hospital births when there was sufficient care, even when the technology of modern obstetrics was only partially available.

If reproduction is conceptualized beyond the biological, as is the case in this book, to include the reproduction of social groups and social life more generally, then "control over reproduction" extends far beyond the provision of technologies and techniques that influence biological fertility within individual bodies. Courts, for example, are institutions within which normative assumptions about family, marriage, comportment, and consent pertaining to socially legitimate reproduction are sustained and enforced. In Vanuatu, as part of the marriage ceremony, it was (and is) common for the groom's family to give the bride's family a "bride price." The gifts that passed between families and kin groups when women married into new kin groups were an object of contention for colonial authorities, who argued that such practices demeaned women and delayed reproduction by requiring prospective husbands to first acquire sufficient means to afford the bride price and ceremony. As a measure to combat low fertility, the British attempted to lower the bride price. At the same time, the Condominium began holding "native courts" in the early twentieth century in order to enforce the colonial laws (collectively referred to as the "Native Code") pertaining to offences between Indigenous people. Ni-Vanuatu, especially women, responded to these courts primarily by bringing marriage disputes and especially divorce cases, which entailed the return of bride price, to the courts.

Reproductive Relations

I clearly take an expansive approach to the subject of reproduction. I see reproduction as including the biological creation and care of new humans, along with, but not limited to, those processes that have been called "social reproduction," namely, the social, political, technical, and economic processes that come to influence the unequal reproduction of bodies, groups, and environments. The episodes I recount in this book show different aspects of reproduction, illuminating the circulation and production of quantification and of institution building in a place where these were first introduced under colonial circumstances. I show, wherever sources allow, Indigenous innovation, affirmation, and rejection of imported reproductive concepts and practices. Though my methods and sources follow the social, political, and epistemic aspects of reproduction, I view reproduction as a natural-cultural or bio-social phenomenon. The sources I analyse there – colonial and postcolonial surveys and documents, oral histories, interviews, and ethnographic engagements – render or produce aspects of reproduction as social; they do not merely make a pre-existing reality visible. These natural-cultural practices have always transgressed – even ignored – the historically European divide between nature and society (Latour 1993). As an imperfect solution, I use the term "reproduction" throughout, sometimes qualified by "social," to remind the reader I am not just referring to the biological aspects of reproduction. This is necessitated by the cultural divisions embedded in English; there is no one term to encompass the interconnections of the social and biological aspects of reproduction that I am concerned with here.

The state institutions and scientific forms of knowledge pertaining to reproduction in Vanuatu that I present here all demand careful consideration of the activities and values associated with relationality and exchange in Vanuatu. These activities, such as "subsistence" – a common placeholder for unpaid care and work with children, food, land, animals, and often including women's labour – were the focus of colonial attention. These attempts to quantify and measure were concerned with controlling, in the name of improvement, not only the physical aspects of reproduction, but its social aspects as well. Quantifying the non-monetary aspects of these processes has been a central aspect of biopolitics in Vanuatu. Studying reproduction in this capacious manner makes possible a critical apprehension of encounters and entanglements among multiple knowledge systems, socialities, and forms of governance.

Reproductive Economies

The interconnection between reproduction, in its biological and social senses, and capitalist economies has been of enduring interest to social theorists. Since the 1970s, Marxist feminist theorists have demonstrated the central place of unpaid labour and care in the reproduction of social life. They further developed the Marxist concept of "social reproduction," which Bhattacharya names as "the activities and institutions that are required for making life, maintaining life, and generationally replacing life" (2020, n.p). Relatedly, the term "reproductive labour" names the care labour, subsistence agriculture, and domestic work (often unpaid) that are crucial for the reproduction of capitalism and its hierarchies (Bhattacharya 2017).

While this political and conceptual terrain has been monumentally important, Murphy (2015a) suggests an even more expansive framing of the already capacious understanding of reproduction in this work. She explains that this is because "in this vein of theorization, reproduction is typically conflated with childbirth and childcare, thereby concentrating the theorization of reproduction on the ways sexed embodied difference has been mobilized in the history of capitalism to create new divisions and stratifications of labour through patriarchy" (287). Reproduction for these theorists matters in relation to the reproduction of labour power. Murphy argues that reproduction should be expanded and considered "distributed" (2013, 2015a, 2017), so as to clarify that it is not limited to the (hetero)sexual reproduction of human bodies or labour power. Reproductive processes require collective and inclusive infrastructures and environments to flourish: reproduction is a process that involves far more than the individual body. Murphy argues that this pushes a feminist politics beyond the scale of the individual body with rights, to where "relations of reproduction are subsumed by but not reducible to relations of production" (2015a, 300). This conceptualization is a way of expanding a Marxist feminist framing of the politics of reproduction as well as refusing what Murphy calls the "economization of life" (2017), wherein the quantification of life, achieved through epistemic infrastructures, is connected to the hierarchical economic valuation of lives through indicators like GDP.

Pacific Islanders have taught generations of anthropologists and anthropology students that reproduction is not limited to the categories of production and reproduction of the critical analysis of capitalist economies. The non-monetary exchanges in Pacific lives cannot be contained by the conceptual terrain of social reproduction, a domain demarcated in relation to (capitalist) production. By sharing information about their lives with researchers, Pacific Islanders have imparted the significance of the connections between reproduction, the exchange of objects, and gendered relationalities. The relationships formed through the exchange of yams, mats, and pigs, for example, are part of the reproduction of bodies, social groups, gender, food gardens, and well-being (e.g., Bamford 2007; Butt 1998; Lind 2014; Malinowski 1916; Strathern 1988; Weiner 1977, 1980). Of particular importance for this book, moral persons, social groups, and connections to land were made and reproduced through exchange relationships mediated by cash, commodities, as well as gifts. As Christopher Gregory has written, the colonial economy imposed on Papua New Guinea did not replace gift economies. Rather, the introduction of commoditized exchange accompanied an "efflorescence of gift exchange" (Gregory [1982] 2015). He argues that both commodity and gift exchange are central aspects of the relations of reproduction, particularly as they are framed through kinship relations that reproduce humans, social groups, objects, land tenure.

These objects, which could also be of European origin – for example, calico (cotton cloth) – were (and are) exchanged at important life cycle ceremonies, seasonal rituals, and, as Lederman (1986) showed, in everyday relations. The division between the biological and social aspects of reproduction cannot be reduced to the reproduction of capitalism and labour power, as immensely powerful as those domains produced by capitalism are. Reproduction in the south-western Pacific Islands (as elsewhere) has always been a relational and distributed process, one that extends beyond the reproduction of individual biological bodies, but anthropologists have not yet examined the colonial, technological, and infrastructural dimensions towards which Murphy's notion of "distributed reproduction" points.

My focus in this book concerns the social, political, and infrastructural entanglements of the quantification and medicalization of reproduction in colonial and postcolonial processes. This entails highlighting Ni-Vanuatu socialities and reproductive economies whenever the empirical sources allow, and in particular when these aspects of life were the focus for researchers and government officials. This means I take up Murphy's challenge in *The Economization of Life* (2017) to consider situated engagements with her global analysis of the connections between reproduction and economics in research and governmental infrastructures. Ni-Vanuatu reproductive economies are present in the divorce disputes that unfolded at the early twentieth century, which I discuss in chapter 1 in the context of colonial-initiated "native courts." They are evident in the way that rural midwives were compensated during the 1960s, as I show in chapter 2, and in the 1967 census categories that render "subsistence" visible in chapter 3. Ni-Vanuatu forms of exchange relationalities are implicated in the emphasis on the marital status of "young mothers" that I analyse in chapter 4, and in the quantification of "alternative indicators of well-being for Melanesia" that speak to indicators that privilege monetary values like GDP, as seen in in chapter 5. The ability to recognize reciprocal relationships that are conducive to well-being is an important aspect of the skills cultivated by several massage healers whom women visit during their pregnancies, as I also show in chapter 5.

There is a great diversity of exchange relationships, including within the expanding purview of capitalism in Vanuatu, and these have changed over the course of the century I consider here. There are, therefore, many aspects and details of Ni-Vanuatu ontologies and economies associated with reproduction that I do not systematically explore in the following chapters, as earlier anthropologists of reproduction and economic anthropologists who focused on village ethnography have done so comprehensively. I do, however, hope to show how colonial and postcolonial attempts to document and influence a population's reproduction were entangled in the Indigenous networks of relationality and economies of care.

Synthesizing Documents, Oral histories, Interviews, and Ethnographic Engagements

Building on ethnographic engagements in Vanuatu that began in 2001, from March to August 2010 I interviewed retired midwives, nurses, massage healers, women who had given birth at the Paton Memorial Hospital, and mothers of children aged less than a year in Pango and Port Vila. I interviewed senior men and women who were respected experts on the history of Pango Village. I attended many public events in Port Vila and Pango, focusing especially on those organized around children's and mother's issues. I had planned to interview mothers of ten-year-old children in 2020, but COVID-19 prevented this, so I continued following these issues of concern online.

Particularly with older women, the interviews became a chance to make something public. When explaining to others why they had agreed to an interview, these women would sometimes say, "My words will go on." In telling their stories here, I am attentive to the stories they wanted to make public and the ethnographic context in which they are embedded. I was and remain less interested in uncovering the previously hidden details of people's lives than in contextualizing what they told me as stories and facts that they explicitly wanted to make public, or that I witnessed in public arenas. For me, this complements the considerations that Simpson (2014) described in regards to ethnographic refusal. Simpson argued that anthropologists need to be concerned with the contours of what people refuse to talk about, since interviewees know how their knowledge might be used against them in multiple nefarious ways. I strove to be attentive and to ask about what people were comfortable telling me, and to present this information as sensitively as I could in terms of what they wanted to make public. My whiteness was an ever-present mediating fact in fieldwork with people who were not. I generally had the sense that discussions with me were not "everyday" encounters in the same way people would have talked with each other in my absence. Whiteness located me in a particular position within a hundred-and-fifty-year history that my interlocutors recognized: whiteness meant the gaze of self-proclaimed expertise, wealth, and, above all, unearned privilege.

While reflexive practices that situate the researcher's identity and position in the process of knowledge production are commonplace in anthropology, anthropologists less often connect research practices and knowledge to labour conditions and social reproduction. As historian Callaci (2020) notes, recognition of the debts of social reproduction that researchers incur are often relegated to the acknowledgments in their formal publications, and they are rarely seen as part of knowledge production itself. The research for this book meant moving from Toronto to Auckland to Vanuatu to Berlin and back to Toronto with small children. Given the immersive realities of fieldwork, and the years of writing at kitchen tables and other improvised locations, my children and partner were compelled to be part of every step of this process. When I was interviewing mothers of children under twelve months, my youngest was just a few months older. She was bounced on laps, her eating habits formed the topic of many conversations, and her fevers were a source of much anxiety on my part. My oldest, who turned five while we were in Pango, attended a nursery school whose generous teacher and her assistants helped us immensely. Researching and writing this book took place over ten years of precarious post-PhD work. This entailed a post-doctoral fellowship, a research fellow position, and then teaching three to four courses per term – summer terms included – for several years. My time as a precarious worker ended, thankfully, in part because of the collective agreement of the part-time faculty union at York University, which allows experienced contract faculty to compete for annually allocated long-term positions. The mixed methodologies adopted in this book, combining archival and ethnographic methods, and the long periods of absence from Vanuatu, were part of my creative response to the realities of precarious labour and raising children. While the research for this book was supported by various official funders (CUPE 3903, York University, and the Max Planck Institute for the History of Science, and earlier research funded by the Wenner-Gren Foundation and the Social Science and Humanities Research Council of Canada), the book has been subsidized by many others in my life.

In terms of the archival dimensions of this book, I take the documentary practices of colonial governance seriously. That is, my intention is to consider not just the content of the archive but also its form (Stoler 2002b). The colonial documents pertaining to quantification, courts, the census, and hospitals produced a public domain subject to reproductive interventions on the part of colonial authorities. The connection drawn between a feminist and an anthropological analysis of reproduction, on the one hand, and an examination of how states "see" through quantification (Scott 1998), on the other, is an important contribution of this book. With a focus on how quantified figures mattered in institutional strategies that made reproduction public in a particular way, the analysis presented here elucidates the social lives of documents and expertise, viewed not just in terms of their content, but especially as modern social forms, as conceived in Matthew Hull's (2012b) *Government of Paper* and Timothy Mitchell's (2001) *Rule of Experts.* The book adds scholarship on the governance of reproduction to this body of literature through its examination of scientific and colonial preoccupations with accumulating population numbers to formulate social policies for the future. The documentary practices mediating the concern with numbers and the categorization of populations were instrumental in binding reproduction and economy together, establishing a connection that would be negotiated repeatedly through encounters between Europeans and Pacific Islanders, as well as between Pacific Islanders.

Documents of various kinds figure prominently in each chapter and are combined with other sources. A principal reason for this methodological approach is that documents are some of the key means by which reproduction is quantified, recorded, and then circulated as a technology of governance. Simply put, documents, their circulation, and their publics were part of my object of study. In chapter 1, I present colonial correspondence chains gathered together in colonial files, comprising a documentary history of how certain areas of life become objects of governance. I adopt a similar approach in chapter 3, where I incorporate published handbooks, demographic research, and census reports as "paper informants." In chapter 2, I combine colonial correspondence with oral histories from midwives and senior women with participant observation conducted in Pango Village. The colonial letters and reports, and the terrible hospital birthing conditions they make public, tell the history of a particular hospital, which I juxtapose with oral histories of fond memories of that hospital. I try to suture together these two seemingly opposing histories by situating them in an ethnographic context. In chapter 4, I present ethnographic research in Pango and Port Vila, along with interviews with women who are mothers of children. I combine this ethnographic research with discursive analysis of national, regional, and global reports and policies on family planning and women's empowerment. I present the texts with an eye to their powerful discursive content and the scales of the publics their discourses produce. The combination of the knowledge derived from ethnographic work and the information contained in the documents shows the local politics that population discourses miss. In particular, the demographic discourses emphasize women's access to wage labour and contraception but fail to address the land access that is a key village concern. Finally, in chapter 5, I discuss the form and content of texts about "alternative indicators of well-being," which I conceive as a type of Melanesian knowledge in tandem with female massage healers' knowledge of well-being that I learned from my interviews and ethnographic research. In combining ethnographic, archival, and documentary analysis, I move between the scales of the peri-urban village, the nation, the region, and the globe. This is to show the multi-scalar infrastructures that shaped reproduction, both in and outside of bodies, in this Pacific Island place and time, and how even lives lived in seemingly remote Melanesian villages are subject to multi-scalar influences, as Harradine (2014, 16) has shown for land issues in Vanuatu. It is also to show that reproduction is intervened on through public discourse, infrastructures, and practices that take place at different scales.

The documents that comprise my historical and ethnographic sources are thus important both for their discursive content and their materiality as they participate in making reproduction public. While I cannot exhaustively examine how they are produced or how they circulate at all scales, each chapter raises at least one way in which these documents come to matter in the quantified representation and governance of reproduction. More generally, in remaining open and reflective in my approach, I am able show that documents do not merely represent social life; they are not secondary, nor are they less material or real than face-to-face encounters. Documents are central to social life itself – and to bureaucratic governance in particular – in addition to constituting the basis for the telling of histories.

Most of the documents I consulted for this book are part of the world making of English-speakers, from British colonial officials, to population scientists and demographers, to Presbyterian and Anglican missionaries, to development organizations and government policy. This connects with my aim of telling histories of the present, at a time when I was situated in Pango, listening to the histories of people who were Presbyterian or members of other Protestant denominations. Some of them had worked or given birth in the country's largest hospital, the Presbyterian Paton Memorial Hospital, which became the government-run Vila Central Hospital. English continues to be the imperial language of many development organizations working in the region. Of course, French colonial governance and French settlers' presence have also been influential in Vanuatu (e.g., Riou 2010) and the region more broadly (e.g., Aldrich 1990, 1993). This means my book is necessarily a partial account of the islands' social circumstances that reflects the English-speaking publics for which reproduction was made visible.[7]

Overview of the Chapters

Each chapter presents a moral figure through which the social, political, and biological processes of reproduction in a given population were made public in Vanuatu. Chapter 1 follows the figure of the imbalanced sex ratio (i.e., considerably more men than women), which was the population figure that stood out as a measurement of poor population health in demographic debates during the period of low fertility in the early twentieth century. The circulation of this moral figure – as a public depiction of a population – by colonial authorities was closely aligned with the desire of Presbyterian missionaries to regulate marriage practices that were based on Indigenous reciprocities necessary for social reproduction. The colonial authorities hoped that lowering what they called the "bride price" to facilitate men's marriages at younger ages would lead to an increase in the population. Such payments at the time of marriage were of central importance for the reproduction of both social relationships and human beings. In this chapter, I argue that the circulation of the figure of the imbalanced sex ratio was important in producing (but not determining) a new social domain subject to colonial policies and moralized interventions related to marriage, sexuality, and consent.

Reducing the bride price met with limited interest among Ni-Vanuatu. When it came to colonial interventions in marriage, the Condominium had better success with its attempts to hold "native courts," declaring that "the shortage of women is the cause of these courts" (Nicol 1935). This new colonial legal forum for conflict resolution signalled a novel way of making reproduction public and visible for governance. Such courts became places for resolving disputes that could not be resolved at the village level. A common kind of case brought before these courts involved women seeking to get divorced. Debates would ensue about the proper cultural practices for determining the husband's family's entitlement to the bride price given at marriage, as well as the role of women's consent in marriage. These courts took the form that they did because of Ni-Vanuatu concerns with the relationships surrounding exchange and reproduction. They also made particular kinds of women's agency visible in ways that aligned with Europeans' interests in ending traditional forms of marriage, and rendered other kinds of agency unimaginable (such as agency within marriages established through a bride price). Moreover, twentieth-century European researchers framed women who deployed these other kinds of agency, using traditional fertility-control methods, for example, as resistant to modernity. In this chapter, which presents material from the period 1910–50, I show how Indigenous concerns associated with the exchanges of reproduction were also economic concerns that could be linked to colonial practices that rendered reproduction public.

In chapter 2, I follow the importance of Ni-Vanuatu nurse-midwives as moral figures who made hospital births possible through their care and expertise, which effectively combined Ni-Vanuatu systems with Western medical knowledge. The period analysed in this chapter marks a shift from the village to the hospital as the most common birth place for women in the peri-urban village of Pango. This transition was instigated by Presbyterian missionaries' attempts to expand their hospital (funded by the British authorities),[8] but, more importantly, it was also prompted by the work of Ni-Vanuatu women who started to train as medical caregivers of birthing women. The nurses and the hospital are remembered fondly in women's oral histories for their care; as some senior Ni-Vanuatu women in Pango told me, "the nurses looked out for us!" This was in spite of the archival evidence that birthing conditions in the hospital were actually dangerous, and the doctor in charge frequently demanded better birthing conditions – now a matter of public concern – from the colonial authorities. Ni-Vanuatu nurses are the moral figures of this chapter, who, as I demonstrate, participated in the medicalizing of reproduction by expanding the clinical capacity of clinics and hospitals. It is commonly held that medicalization is a process that brackets off social and political worlds as it individualizes biological life processes. What I show here is that attempts to medicalize childbirth were also constitutive of a process that brought it into the public realm, where medical staff could advocate for safe birthing conditions as a matter of public concern to the colonial government.

These attempts to medicalize birth could not have happened in the absence of Ni-Vanuatu knowledge, technologies, and economies of care, which were, as I call them, relational infrastructures. In contrast to Ni-Vanuatu men, who were selected to train as assistant physicians because of their Western education, Ni-Vanuatu women were chosen to train as nurses because of their knowledge of village birth practices and their involvement in village networks of reciprocity. Even the colonial administration's system of paying village midwives, who were incorporated into the formal medical system, emphasized Indigenous forms of exchange and care. These networks of reciprocity and care around birthing women and their birth attendants, as well as their skills with technologies – biomedical and otherwise – demonstrate the entanglement of the social, biological, and technological aspects of care for reproduction in Vanuatu. Mol, Moser, and Pols (2010, 14) argue that care is "persistent tinkering in a world full of complex ambivalence and shifting tensions" that resolutely involves technologies that themselves work through care. The recruitment, training, and work of the nurses demonstrate colonial dimensions of care where tensions concern forms of knowledge that are not held to be equal and technologies that are scarce for structural and racist reasons.

The participation of Ni-Vanuatu in the provision of biomedical care contributed to the formation of Indigenous modernities, providing opportunities for professionalization and for building up infrastructure that would eventually be taken over by the national government. To the extent that the hospital was successful, it entailed women's involvement, and their care, in the institutions that would become part of the nation-building process. Senior Pango women's fond memories of the nurses, which were generally juxtaposed against the neglect experienced at the 2010 government hospital, can be read as an expression of postcolonial discontent over the fact that modernity's material promises failed to materialize.

In chapter 3, I analyse the paper trails leading to the 1967 census, the first undertaken in Vanuatu. Norma McArthur, an Australian demographer, and J.F. Yaxley, a British administrator, undertook the project of training Ni-Vanuatu to be enumerators across the archipelago. They instructed the census takers to justify the census by claiming that "it will help planning in the future" (McArthur 1967b). The decline in population was thought to be in reverse, and accurate figures were needed to develop public policy and institutions. I argue that census categories and practices, and especially the figure of "subsistence," transformed social reproduction into public knowledge but separated it from land politics. "Subsistence" here refers to the agricultural work needed for food, shelter, ceremonial exchanges, and kin obligations. This census tracked the number of births per woman, basic household sizes, and, significantly, paid and unpaid livelihoods. Thus, in the context of this first full enumeration of the population, I illuminate the challenges faced by the planners in their attempts to accommodate Indigenous social forms into their categorization of "subsistence" within census questionnaires, connected as it was to exchanges needed for social reproduction. "Subsistence" emerges as a historically specific knowledge category that was aimed at quantifying life and making the non-monetized reproductive labour of women and men publicly visible in particular ways. In fact, the inclusion of "subsistence" on the census rendered women almost as economically active as men. The chapter shows that the public quantification of life within the census rendered men's and women's work and care legible to state and development agendas, conscripting them as subjects to be planned for in particular ways. Local forms of production and exchange, especially around copra production (glossed as "subsistence") influenced the forms that this knowledge took. Access to land and changes in land use associated with settlers' changing agricultural practices were discussed in different fora, enrolling Ni-Vanuatu in a state politics that presumed and planned for wage labour as an issue that was separate from the question of land access.

In chapter 4, I show how the figure of the "young mother" developed as a social identity and a demographic category in ways that connected particular economic concerns to reproduction at the national, regional, and global scales. In the daily life of the village, the figure of the "young mother" evoked both pity and blame for her apparent failure to think about the future. By contrast, the term "young father" lacked the same social traction. Through ethnographic research, I show how the public narratives about "young mothers" contributed to the social marginalization of young women, especially during pregnancy. This situation created conditions poignantly expressed by one such young woman: "I just wanted to be invisible." Anxieties about the right time to become a mother are also reflective of the debates that ensued regarding new options for women, new pressures on wage labour, and gendered tensions about who could work in town and gain access to land through marriage. I argue that the public figure of the "young mother" consolidates contemporary anxieties about waged livelihood as well access to land in the village.

In national, regional, and global population discourse, young women's fertility is economized and even financialized in order to achieve a good future. Population reports and policies make women's ability to control fertility through biomedical contraception ever more crucial for their access to education and wage labour. The concern with "young mothers" at various scales reveals the entanglement of reproduction with wage and traditional economies. Young women's experiences show how wage labour and land-based subsistence strategies create particular power relations during times of population growth and, as McDonnell (2016) shows, shifts in land use practices that favour men's access and undermine women's access. These complexities of the relations of reproduction cannot be accounted for in demographic and population policy discourses' concern with women's economic empowerment, wage labour, and biomedical contraception at larger scales of reproduction.

In chapter 5, I present the quantified figure of "well-being for Melanesia" as derived from the collection of data described as "alternative indicators." I show how it renders Ni-Vanuatu values about social reproduction public. This figure is a quantified category that the statistical experts within the Ni-Vanuatu government developed as a counterpoint to GDP and other financial indicators of well-being, signifying a stance of sovereignty on a capitalist frontier. This figure, which is purposely quantified to serve as the foundation of evidence-based governance, focuses on exchange practices associated with food, shelter, and important life events.

And yet, the well-intentioned figure of "well-being for Melanesia" underplays important aspects of local care and well-being, especially for women and children. I explore the knowledge and spaces of care that massage experts provide for most pregnant women and young children. In addition to the knowledge entailed in the manual manipulation of massage, these women assert their expertise through their knowledge of the unequal relationships that invite ill health. In short, just like the quantitative figure "well-being for Melanesia," these women know how local lives should be lived, but they also emphasize the critical importance of care and reciprocity. They are moral figures at a different scale. Taken together, the respective knowledges of the public figure and local healers demonstrate that well-being is a way of asserting how lives can be well lived on a postcolonial capitalist frontier.

Finally, in the epilogue, I reflect on the Ni-Vanuatu futurities implicit in each chapter in the light of the recovery process that followed the category-five cyclone that destroyed food and material infrastructures in northern Vanuatu in April 2020, a time when the COVID-19 pandemic had already closed borders everywhere. For, as Murphy (2015a, 287) writes, the question "how might we build new forms of life out of the old?" should accompany the theorizing of reproduction. While many aspects of reproduction have been made public in order to govern and monetize many aspects of our existence, much life still exceeds and pushes past efforts to regulate its reproduction.

Chapter One

"The Shortage of Women Is the Cause of These Courts": Imbalanced Sex Ratios, Native Courts, and Marriage Disputes Made Public, 1910–1950

My attention has often been drawn back to a photograph of two women from the island of Tanna, whether reprinted on Vanuatu Cultural Centre posters or in the book for European popular audiences by Felix Speiser (1913, 272). The women have often animated my thinking about this time period in Vanuatu because their expressions counter the feelings of despair engendered by researchers' public retelling of women's statements during years of social upheaval in the late nineteenth and early twentieth centuries, when infectious diseases decimated many villages. "Why should we go on having children? Since the white man came, they all die" is what several researchers reported women saying during this period. In the photograph of the two Tannese women (see figure 1.1), one looks directly into the camera, and I read her expression as conveying a powerful presence of survival, endurance, and resolve. The other woman looks away slightly, her brow somewhat incredulous and her lip slightly curled, and I imagine her thinking about Speiser's awkwardness in taking the photo: "What, exactly, are you doing?" Or perhaps her expression conveys a response to something unrelated to Speiser and his camera.

It is not possible for viewers to know what these women were thinking or feeling when the photograph was taken. I open this chapter with the photo as a reminder of the many forms of resistance and refusal in Vanuatu: Ni-Vanuatu women may have steadfastly returned the gaze of the colonizer or looked slightly askance with irritation. I kept these images in mind because they interrupt colonial depopulation narratives that emphasize the powerlessness and desperation of Ni-Vanuatu men and women. It is important to reiterate here, in the face of the colonial documents discussed below, that the complexities of Ni-Vanuatu life worlds always exceeded attempts, colonial or otherwise, to write about them. I invite you to read these colonial scripts

Figure 1.1. Women from Tanna
Source: Speiser (1913, 27).

and to unpack colonial thinking and actions with me so that we may understand the past and the present differently, as well as achieve multiple readings of women's agency and actions, especially as they relate to reproduction.

Whereas diseases brought by Europeans were generally understood to be the cause of the high mortality rates in the early twentieth century, low fertility rates were harder for the researchers to explain, though this did not stop them from blaming Ni-Vanuatu women's "insouciant" or "careless" attitude towards mothering (Jolly 1998, 183). This was an early twentieth-century instance of the still common racist trope of the "bad mother" in population discourses. Ni-Vanuatu women's fertility control in a context of adults' and children's high mortality rates from infectious diseases demonstrated, in the European scholarly imagination, a lack of hope for the future. In their publications and reports, researchers attempted to shed further light on these high mortality and low birth rates. While they did not understand the low fertility rates, researchers at the time were certain that these lives were tenuous and in need of care, or at least protection. I focus on colonial efforts aimed at increasing the number of Ni-Vanuatu births to show how a wide constellation of social and biological concerns was constituted into a set of issues by which reproduction became governable.

The imbalanced sex ratio was a demographic figure that particularly stood out in scientific debates about fertility and survival at the time: "The shortage of women is perhaps the most ominous sign of decadence, for it influences the fertility of the race for generations to come" (Speiser [1923] 1996, 50). British colonial authorities, working in conjunction with Presbyterian missionaries, sought to understand different marriage practices in order to develop a policy that would lower the "bride price," as it was called in colonial documents, in an attempt to increase fertility. In doing so, they frequently referred to the sex ratio to justify this policy. In this chapter, I follow this demographic figure and its significance in relation to the overall strategy of colonial governance in the newly formed polity of the New Hebrides.

In *The Economization of Life*, Murphy (2017) argues that the 1920s saw the beginning of a significant shift whereby the reproduction of populations was understood as something that could be controlled for economic concerns of a national economy. This would be different from the Malthusian connection between reproduction and political economy, where population growth would eventually lead to famine, to one where population measurement could be optimized. That a colonial power would invoke a sex ratio at this time, a population figure, in the context of an intervention on reproduction is an important historical development, but it also met with local realities and concerns.

My methodological and analytical choice is to begin with and follow a demographic measurement and its entanglement in social policies and practices. I could have framed this history through a narrative constructed around how British colonial officials attempted to control violence within and between families, or how missionaries sought to promote monogamous marriages. By instead tracing the figure of the sex ratio in colonial documents, files, and paper trails, I show how quantified representations of population circulated in colonial networks at a time when reproduction was made public in new ways. At this historical juncture, this process was entangled with the production of a pan-village public polity and with attempts to expand colonial governance in the New Hebrides. I argue that the circulation of this demographic figure contributed to (but did not determine) a new social domain subject to colonial policy and moralized interventions on marriage, sexuality, and women's consent. Furthermore, it was the exchange practices relating to marriage, or bride price, that facilitated this public sphere.[1]

Understanding the colonial entanglements of population figures means taking the documentary *practices* of colonial governance seriously. Generating documents and figures can represent an attempt to accomplish, if not achieve, administrative control while constructing particular subjects, objects, and socialities (Hull 2012a, 253). In highlighting the socialities that these documents and population figures produced and rendered visible, I show how they played a role in producing a public domain, one in which kinship was subject to colonial interventions in the New Hebrides that worked hand in glove with the missionary enterprise of seeking to enforce particular kinds of domestic spheres

(Jolly 1991). The production of this public domain demonstrates how reproductive and economic concerns were interwoven within colonial and Ni-Vanuatu worlds. While colonial officials and missionaries disagreed with the practice of the bride price because it meant, in their eyes, that women were considered equivalent to pigs, they were actually adept at rendering this marriage practice in economic terms, and thus amenable to intervention. For Ni-Vanuatu, this colonial intervention aimed at controlling "the price of brides" accompanied the increasing circulation of cash within their exchange networks/economies: bride prices were high because men had access to cash wages from labour contracts on plantations outside of their villages[2] and, later, because copra[3] prices were high. The changes wrought by the presence of cash did not entail a completely new form of exchange; indeed, Jolly (2015) argues that the colonial economy took the form that it did because certain Indigenous objects were already fungible. Christopher Gregory (2015) shows that a colonial economy of imposed commodified relations in Papua New Guinea by no means replaced non-commodified exchange. Furthermore, he argues that the exchange of the bride price is key to understanding the reproduction of persons and social groups. This chapter shows how colonial documentary mechanisms contributed to the production of a public domain associated with kinship and exchange.

The development of state projects has been connected with the separation of the public and private domains. Whereas the state appears to be associated with public and economic affairs, and not, it would seem, with intimate matters like kinship, in reality, as Lambek argues, "kinship is embedded in the fundamental actions of the state … The state is constituted in and through such acts as … providing birth and death certificates … and more generally producing and authorizing the means by which people are related to one another as parents, offspring, spouses, siblings and the like" (2013, 26). The regulation of marriage in particular figures prominently in relation to the values and practices of statecraft. McKinnon writes that in United States during the second half of the nineteenth century, the stigmatization of polygamy and cousin marriage accompanied the political promotion of "equality, individualism, choice, and secular contract law central to post-revolutionary ideals of political governance, not to mention 'free' economic labor" (2019, 612). Colonial states often concerned themselves with kinship matters in multiple ways, a focus exemplified in their regulation of "mixed marriage" in the Dutch Indies (Stoler 2002a), genealogical thinking and bilateral descent in colonial Kenya (Holmes 2009), and mothering (Manderson 1996; Hunt 1997; Coghe 2022). Here, I explore the entanglement of population quantification and colonial policies and interventions in marriage and divorce in the New Hebrides. Authorities in Vanuatu were concerned with two areas in Indigenous marriage: bride price and reportedly high levels of violence between intimate partners, as well as between families. The policy aimed at regulating the bride price and reducing violence before the dissolution of marriage focused on exchange – and on what should not be exchanged. On the one hand, colonial modernity in the New Hebrides meant stigmatizing bride price because it obfuscated the separation of the domestic and economic spheres. On the other, the economic logic of supply and demand, and the fact that women were in "short supply," made it thinkable for colonial authorities and local chiefs[4] to sometimes work together to attempt to regulate marriage through the mechanism of the bride price. I further show that the superimposing of a domain subject to colonial policy, wherein colonial officials valorized marriages that entailed women choosing their own husbands and not being "sold," made certain forms of female agency visible.

I begin by showing how the figure of the imbalanced sex ratio rendered social and biological processes in New Hebrides visible to particular publics outside of the archipelago. In the remainder of this chapter, I discuss how the sex ratio circulated in a local context with colonial intentions to increase fertility, regulate marriage, and reduce violence in the New Hebrides, starting with the southern part of the archipelago and progressing to the northern part. In so doing, I examine colonial attempts to reduce the bride price and the development of "native courts" to resolve conflicts, which were often about adultery and the return of the bride price at the termination of a marriage. I conclude with a discussion of forms of women's agency made public in colonial correspondence and policies. Throughout this discussion, I show how colonial efforts connected reproductive and economic concerns and how demographic knowledge informed attempts to expand colonial state forms of governance that made reproduction visible as a public concern on a colonial frontier.

Circulating Sex Ratios and Their Publics

In the early days of global population thinking, the sex-ratio figures of the New Hebrides circulated outside of the archipelago within cosmopolitan research networks and entered debates about population health and racial degeneration (Widmer 2014). It was the imbalanced sex ratio that made the population politics of the region visible at the landmark World Population Conference in 1927 (Sanger 1927, 247). Also at that time, the sex ratio circulated in networks of people critical of colonial labour practices and the treatment of Indigenous peoples. Concerned citizens in Australia, such as the members of the Aborigine Protection Society, who commissioned a report by Speiser in 1920, could read about the sex ratio in conjunction with labour recruitment practices. Gendered recruitment practices that had begun prior to the implementation of the so-called White Australia policy continued after the labour migration of Pacific Islanders to Australia came to an end in 1901. After 1901, men left their homes to work on plantations in New Caledonia, or in other parts of the New Hebrides, while women stayed in their villages. Whether women should be allowed to work for wages away from their homes was a subject of debate among French and British authorities in the New Hebrides. The former needed to respond to the complaints of French settlers (significantly outnumbering British settlers), who wanted labourers and were considering hiring New Hebridean women for additional jobs other than domestic service. British officials were more responsive to missionaries' claims that women should remain in the home. Labour migration would result in a loss of women's labour and reproductive capacity within the villages, as evidenced by the low ratio of women to men. Whereas the low numbers of women might have justified the need for women to remain in the villages to increase the population, missionaries and Indigenous men shared a concern with controlling women's sexuality, which they achieved by limiting women's migration possibilities (Jolly 1987). Still, what researchers like Speiser and Rivers called for was an end to recruiting in general, so that men could stay in the villages or work on nearby plantations. The unbalanced sex ratio constituted numeric evidence in their critique of labour migration and the unjust treatment of Pacific Islanders, as well as the naturalization of their place in their villages.

Within the New Hebrides, the imbalanced sex ratio was of great local significance. It was said to be as high as 159:100 on the island of Espiritu Santo (Baker 1928, 282), though this was an imprecise calculation at best, given the very limited availability of population statistics. Therefore, this figure tended to circulate in a narrative form as "the imbalanced sex ratio" rather than as the actual number. During the period covered in this chapter (1910–50), the New Hebrides was a polity still in formation and a colonial outpost of the British and French Empires where a small administrative colonial settlement, Port Vila, had been established along with district offices on the islands of Tanna, Santo, and Malakula. The archipelago was then home to approximately 65,000 Indigenous people speaking over 110 languages, and it was a destination of Presbyterian and Anglican missionaries, Oxbridge researchers like W.H.R. Rivers and John Randall Baker, a handful of European settlers (approximately 25 British and 100 French residents in 1906; Rodman 2001, 33), and Vietnamese indentured labourers (6,000 in 1920; Meyerhoff 2002, 47). There were elaborate political systems at the village level and inter-island relationships, but there was no pre-existing inter-island state polity. Forms of kinship varied across the islands. Wages from labouring on plantations (mainly in New Caledonia and the New Hebrides) were important, but Ni-Vanuatu primarily organized their socialities and land use around the subsistence production of root crops and small-scale animal husbandry, and they built their homes largely from materials available in their environment. Mass media did not exist; the closest equivalent was a quarterly newsletter published by Presbyterian missionaries, mainly for the faithful back in Australia, Canada, New Zealand, and Scotland.

From the Files: "Vital Statistics Land etc." and "The Price of Brides"

After attempting to conduct a census of the island of Ambrym in northern New Hebrides, Dr. Maurice Frater, a Presbyterian missionary, wrote a letter on 27 September 1916 to the British resident commissioner (BRC), Merton King, who placed the letter in the file entitled "Vital Statistics Land etc." Dr. Frater (1916) claimed that the "most disconcerting feature of the census" was the higher number of men compared to women. In his estimation, two cultural factors were implicated in the sex imbalance and associated population decline: "1. Early, immature marriage on the part of the female. 2. The system of women purchase is not now adapted for the propagation of the race … The price of a woman on Ambrym ranges from 10–15 pigs." This "evil is further aggravated by old men having a plurality of wives, many of them, young women who, with husbands of a like age, would be doing their part in carrying on the continuity of the race." Dr. Frater recommended that the problem be rectified by the BRC and strongly suggested that the chiefs forbid the marriage of young girls. The marriage practices were "the evil which threatens to drain the life blood of the native race." In the monograph he based on the fieldwork he conducted from 1910 to 1912, Speiser ([1923] 1996, 40) also stated that the shortage of women implied that "only wealthy, that is, old men can afford to buy wives." He was cognizant of the reality that not all reproductive sex takes place within marriage (40), but as extramarital encounters were irregular (or so he thought), Speiser believed that the chances for conception were higher in a marriage.

Whereas Speiser contended that the colonial authorities had no influence on people whatsoever, Dr. Frater, as seen above, demanded more action. Frater specifically recommended that the Condominium regulate the age of marriage by lowering the bride price because this was a legal strategy that interested the missionaries for reasons that extended beyond merely halting depopulation. The Presbyterians wanted to regulate the bride price because they saw the practice as degrading to women, as it appeared to equate them with pigs and gave them no say in the selection of their husbands. The missionaries also wanted to reshape kinship relations in favour of nuclear families (Jolly 1991, 28), and bride prices were imbricated in the building of alliances between villages and larger kin groups. Reducing the bride price was part of a broader Christian project of imposing middle-class European forms of domesticity (Jolly 1991) and supposedly halting the mistreatment of women.

"Bride price" is an English phrase that should be read as a placeholder during this period because even the colonial authorities would frequently use quotation marks around it and associated words. They would write, for example, of Indigenous marriage practices involving women being "sold," which perhaps indicates their awareness that these practices did not quite correspond to their English equivalents. This exchange of wealth has long been a topic of interest within the anthropology of kinship, feminist anthropology, and Pacific anthropology. Jolly (2015) recounts how the anthropologists Felix Speiser and John Layard depicted exchanges at the time of marriage based on their observations during fieldwork conducted during the periods 1910–12 and 1914, respectively. In light of his travels throughout the archipelago, Speiser wrote that "the purchase price is, so to speak, the rent which the husband pays to a woman's clan for enjoying her favours and making use of her labour" (quoted in Jolly 2015, 68). Immersed in one place, Layard wrote of the "permanent state of indebtedness in which a man stands towards his wives' parents" (quoted in Jolly 2015, 69). Thus, Jolly argues that while these anthropologists were, to varying degrees, sensitive to "collective relations of clanship and cycles of intergenerational exchange" (69), the language of commoditization pervades their texts. Jolly further argues that in recent years, anthropologists have been more concerned with showing how the commodity-oriented language of "bride price" distorts Indigenous forms of exchange. It appears that anthropologists prefer to critique the limitations of a Western world view in order to develop an understanding of exchange outside of its commodity form. Jolly argues that marriage exchanges in the New Hebrides and Vanuatu show characteristics of both the commodity and gift forms of exchange.

Accumulating pigs was important for status and power, particularly in the northern part of the New Hebrides. This was of interest to Baker

(1928, 114), who remarked that one way in which men accumulated pigs was by obtaining them from the grooms at their daughters' marriages. Usually, men could not raise enough pigs by themselves and needed to be "popular" in order to "borrow" some (114). Accumulating pigs did not mean individually raising them or purchasing them with cash; rather, it entailed developing broad social networks that were both individual- and kin-based to assist in this task. The bridegroom would have accumulated the pigs via his social relationships, both personally and with the help of his family. Thus, gaining access to the number of pigs required to raise a bride price was a lengthy process that was only accomplished when men had reached middle or old age.

The shorthand "bride price" entailed acquiring the necessary objects or animals and presenting them at ceremonies, where the exchanges were witnessed by others. In this way, the exchanges produced the identities of those people directly involved as well as the relationships between their families and future children. Among the many relationships reproduced through these bride price exchanges, the ceremonies would clarify relationships around the couple and their children in a way that would be meaningful for inheritance and access to land. At a time characterized by depopulation, when reproduction and social reproduction were at stake, such exchanges assumed heightened significance.

Nearly two decades after Dr. Frater sent his missive lamenting the sex ratio and requesting the regulation of bride price, colonial correspondence revealed the revival of the topics of population measurements and cultural practices. During his brief tenure as acting British district agent (BDA) on Malakula in 1935, Tom Harrison (1911–76), a self-taught anthropologist who undertook expeditionary research in Vanuatu and later in Sarawak, where he became the director of the state museum, had attempted to regulate the bride price, with poor results.[5] Even still, correspondence in the British file titled "The Price of Brides" reveals that attempts to implement the policy were renewed almost a decade later, in 1944. BRC Blandy began by asking the BDA in the northern region how the policy was working. In a memorandum dated 28 August 1944, he stated that after talking with the Presbyterians, he had decided to order the BDAs to continue to use their influence to keep the bride price low:

> While realizing that there may be arguments against this policy, as well as for it, the overruling argument in its favour is to me that high bride prices militate against early marriages of young people and in favour of the acquisition of young girls by old (unfertile) men. It is therefore definitely against the increase of the native population which is so desirable. (Blandy 1944)

The BDA wrote back to say that there had been little success for years in regulating the bride price. Bride prices in the "bush" (a common term used to describe places where the inhabitants had not accepted missionaries or colonial control), where women were exchanged for pigs (and no cash), ranged from 10–20 pigs, whereas in the Bushmen's Bay area (which, despite its name, was a coastal community with colonial contact) and in small islands in its vicinity, the price had remained more stable at A£10[6] and 2–3 pigs (BDA 1945a). In his quarterly report dated April–June 1945, the BDA mentioned that three marriages were brought to his attention; for two of them, the bride prices were A£10 and A£5, respectively. He continued: "Many other marriages have undoubtedly taken place during this period but no complaint of excessive prices has been made to me" (BDA 1945c). He frequently made such comments in his quarterly reports, writing, for example, "I have no doubt that in certain cases high prices are still being asked and obtained, but no complaints made" (BDA 1945b).

From 1945 to 1950, the BDA reported meeting with chiefs at least once a year to try to encourage a reduction in bride price, but with little success. An exception occurred in 1950, when BDA Crozier managed to meet with four chiefs respectively from Atchin and Vao and from small islands in their vicinity to discuss the fact that the prevailing bride price on these small islands off Malakula was as high as A£100 plus "any number of pigs." The BRC read the reports with interest and reiterated that the BDA should persist in trying to meet with chiefs to lower the price. He closed his letter by returning to the question of an economic logic and the imbalance in the sex ratio. He asked whether this high price on small islands was because "marriageable females in small islands are in short supply" (Flaxman 1950). BDA Crozier suggested that bride prices could be this high because Ni-Vanuatu had a lot of cash as a result of high copra prices. It was agreed that the bride price should be reduced to A£50 and one tusker pig (a pig that is especially valuable for its developed tusks). It was important, in the eyes of the BDA, that the price was fixed because there was a fair amount of inter-island marriage, and if the price varied, "fathers of daughters of marriageable age would try to sell in the best market and thus help to depopulate their own island" (Crozier 1950).

With the exception of Seventh Day Adventists, whose marriages involved a feast paid for by the groom instead of a bride price, the content of colonial correspondence relating to attempts to lower the bride price does not include any mention of opposition – either by women or by men – to exchanges associated with marriage. As I discuss in the next section, conflicts about the nature of the bride price became visible to the Condominium when marriages were ending and some parties to the marriage did not get the results they wanted at the village level.

The Value of Women and Native Courts in the Southern Region of Vanuatu

In addition to examining the "Price of Brides" file in depth, I continued to follow the imbalanced sex ratio and concerns about reproduction within the colonial files and their framing of social life. This led me to the colonial and missionary concern with exchanges at both the beginning and termination of marriages, and to the Condominium's dispute-resolution mechanism, described as "native courts,"[7] that were used by Ni-Vanuatu. BDA James Nicol was in charge of implementing the courts in the southern region (on the islands of Tanna, Erromango, and Aneityum). Such courts were intended as a mechanism for resolving disputes and crimes between Ni-Vanuatu and based on the "Native Code."[8]

The Native Code was an instrument for dealing with offences committed by Ni-Vanuatu against other Ni-Vanuatu that allowed the native courts to take Indigenous customs into account when deciding the kinds of punishments to impose. The Condominium attempted to first establish these courts in the southern region, as colonial officials believed that the area's inhabitants were comparatively more receptive to European social forms. Native courts meant that a district agent (DA) would travel to appointed places and adjudicate cases in consultation with unpaid representatives of villages on the island, called assessors and chosen by the local inhabitants. The administration's plan was to engage an equal number of Christian[9] assessors and assessors who had not attended school. Ideally, there would be four assessors (though in practice there were usually two). In particular, if the accused had never attended school, at least two assessors who had not attended school had to be selected. The assessor's role was solely consultative, specifically to help the DA interpret the testimonies. Nevertheless, the assessor was an important individual who served as an expert on relevant Indigenous practices that the DAs often did not fully understand. The DA carried out the sentencing; sentences involved hard labour, often to establish roads, although possible sentences could be prison time or fines paid to the government agent. The administration intended the courts to be places of "conciliation," and if one of the parties disagreed, appeals could be directed to the resident commissioners.

In a letter addressed "to the Chiefs and natives of Tanna," both the British and the French resident commissioners indicated that the courts would deal with crimes of "violence connected with the stealing and abduction of betrothed and married women and violation of the customs of the land" (Mahaffy and Repiquet 1912). The colonial protocols did not define "native marriage"; they merely referred to it. This presented an administrative challenge, as marriage practices varied across the islands and could also be interpreted differently within particular communities. "Native marriage" was only defined according to local custom. No colonial legislation existed for codifying Indigenous marriages, so the colonial approach, as with attempts to lower the bride price in the northern region, was to work together with chiefs and use persuasive tactics rather than imposing European laws through force. Though the courts charged themselves with implementing the Native Code, in practice, Ni-Vanuatu would bring to the court grievances that could not be resolved in their villages. Some of these cases raised challenges because there were based on disagreements over the nature of customary practices between Ni-Vanuatu, between planters and administrators, and between missionaries and administrators.

The Condominium consulted with F.E. Wallace, a lawyer from outside the New Hebrides, to elicit his opinion on the efficacy of the native courts. He pointed out that adultery was by far the most common reason for holding such a court, but that unmarried people were also convicted of this charge. This was in violation of the code, which specified that adultery related only to married people. Recognizing that this definition did not cover the range of situations that would be brought before the native courts, the BRC proposed an amendment in 1930 that would add "seduction," a charge applied to men who had intercourse with both unmarried and married (to someone else) women, to the Native Code. The perpetrator, if found guilty, would pay a fine to the woman's parents. The fine would be higher if the woman was a virgin (proof would be needed), because this would result in her "depreciated value on the marriage market" (Blandy 1930). Effectively, this meant that adultery would be punished by time served in prison, whereas seduction would be punished with a fine. The fines were intended to be aligned with "native customs" as they related to such violations.

In 1930, DA James Nicol responded to the commissioner's concern regarding women's "depreciated value" as follows: "Women are in the minority and are therefore never a drug on the market" (Nicol 1930). By using this phrase (which at the time usually referred to objects that could be sold quickly and easily) to denote supply and demand, Nicol made the claim that women could get married regardless of their sexual history before marriage. Nicol concluded with a statement that the amendment for seduction (punishable with a fine) would never be used because there was no word in the language (it is unclear which of the Tanna languages he meant) for "chastity." Furthermore, he reasoned that "copulation doesn't detract from the value of a female as a wife," so the fine would not be a deterrent. The colonial administrators, who claimed that they were following Indigenous customs, were also policing monogamy and applying an economic logic to the exchange of women. The sex ratio thus became evidence of – and was interpreted in alignment with – the economic logic of supply and demand in the eyes of the men attempting to regulate marriage and sexuality in the name of reducing violence.

In January 1934, the BRC began questioning Nicol about how people in the Southern District were responding to the courts. Nicol replied that they came frequently to the courts, the most common reason again being adultery, which was a criminal offence at that time. The punishments for adultery ranged from A£3 to ninety days' imprisonment with hard labour or six months of work on the roads. The following year, Nicol wrote, "Women come freely to the court in cases of assault or threats to make them marry a man whom they do not want. Every woman or widow or maid has a guardian (male). *The shortage of women is the cause of these courts*" (Nicol 1935; emphasis added).

As for civil matters, the "exchange" and "ownership" of women and girls also figured in Nicol's list of reasons for holding courts of conciliation, right behind the matter of the Seventh Day Adventist converts (vegetarians who had given up raising pigs) refusing to fix their portions of communal fences to protect gardens from pigs (Nicol 1935). Civil cases were handled via arbitration and conciliation, given the absence of any colonial guidelines. In his reports to the BRC, Nicol would repeat statements of this nature. For example, in 1934, he stated that people "avail themselves of the courts on every occasion that they consider they have a grievance, also about many things that do not really concern the courts. Women come freely in cases of assault or trouble over the exchange of women and settle themselves on the premises until the matter is settled" (Nicol 1934).

Interventions in Marriage Troubles in the Northern Region

Further tracing colonial interventions into (social) reproduction in the northern region, I requested and read the correspondence in a file labelled "Marriages by Native Customs and Christian." "Divorce" had been scrawled on the file at a later date. This file could have been labelled "Disputes Arising due to Disagreements about Exchange," or

"Women's Free Will and Consent in Marriage or Sexual Contact," as these were colonial concerns at the time. However, it was marriage and divorce that bound together the chains of letters contained within the file. Colonial attention to marriage and the end of marriages centred on a concern with determining whether customary practices were being upheld, economic contracts fulfilled, and the desires of brides or wives met. The correspondence chains in the file reveal that economic contracts relating to marriage reproduced multiple relationships, that custom was contested, and that free will was a premise that was selectively invoked.

On the whole, because Ni-Vanuatu rejected colonial interventions on bride price (indicated by the general failure of colonial attempts to lower the bride price), marriages were not brought to the BDA's attention until they were over. The BDA (1939) described the end of a marriage as follows:

> When the woman runs away in which case the matter is usually settled by the husband being persuaded to accept his pigs back and thereby liberate the woman, but this question of pigs is most complicated and sometimes involves not only the husband, his father and relatives, but the whole village which accounts for the refusal of the husband sometimes of accepting the return of his pigs the repartition of which would greatly embarrass him. This also applies to the bride's relatives who refuse to hand over the pigs received by them, in which case the woman remains the husband's property as per native customs.

In this respect, questions about how to handle conflicts over the return bride prices and how to mediate peace constituted a common thread in the cases included in the file. I outline two such cases here.

In 1938, Paton, a Presbyterian missionary, wrote repeatedly to BRC George Joy, requesting him to overrule a BDA's decision. The missionary claimed that this was a "test case" of how the colonial administration would align itself because it appeared to require the colonial official to side with either a woman wishing to leave a marriage (and therefore break agreements sealed through marriage exchanges) or with those asserting traditional marriage practices. Paton intervened on behalf of a particular man, Maki-run, who lived in a village on the island of Ambrym. In this case, a "deposit pig," as the official called it, had been given to a young girl's family as a first step in a marriage transaction that would be completed when the girl, Bata, was of proper marriageable age. Now a grown woman, and two years into living with Maki-run, Bata returned to her mother "of her own free will," claiming that her husband had mistreated her but also that the marriage had never been properly completed with sufficient payments. BRC Joy considered the missionary's requests to intervene in the case on the husband's side, but he ultimately decided that custom had been followed. In the words of the BRC, "Maki-run had not carried out his obligations in connexion with the purchase of Bata and in consequence had no claims on her" (Joy 1938a).

There were several reasons for Paton's appeals. The involved parties disagreed about the correct way to handle the "deposit pig." People in the village refused to return it, but the Ni-Vanuatu assessors and some of the chiefs told Paton that the pig should be returned according to custom. To the BRC, it made logical and economic sense that a family that had accepted a pig in exchange for a young girl could refuse to return it. He wrote,

> In some areas of the group [meaning the islands of the New Hebrides], the pig deposited with the parents of an immature female child is named in marriage to a native on attaining maturity, is not returnable should the payment for the woman not be completed. This is reasonable because once the child is bespoken, other and perhaps more suitable suitors may pass her by. The forfeiture of deposit seems just and coincides with civil law. I am not "au courant" with native custom on Ambrym, but it would not surprise me if the same custom prevailed in this respect. (Joy 1938b)

The missionary persisted in asking for help in support of the abandoned husband, saying that Maki-run had tried to meet his obligations by providing the remainder of the bride price, but that Bata's family had refused to fix a date for the ceremony. The BDA, who was not convinced by this claim, stated that two years was a sufficiently long time to fulfil this obligation. Paton concluded by questioning whether the woman was indeed leaving of her own "free will." The men of her village had a fierce reputation, and three previous native courts had already ruled against them. The missionary declared that she had not really left Maki-Run on her own volition, and he had not even mistreated her. Rather, she left because her male family members did not approve of the marriage. The BRC maintained his ruling in favour of the young woman, which would ultimately lead to the least amount of violence within the marriage and between kin groups; it would also entail the easiest path of enforcement, given the reputation of her male relatives.

In a concurrent case from 1937, B.C. Ballard, a lawyer at the Joint Court,[10] wrote to BDA Adams requesting his intervention on behalf of a man whose wife was living with another man on Ambrym. Maiwot, the husband, had "paid 25 pigs and A£4.10 for a woman named Lilon." The husband went to work in Vila, and while he was away, another man, Bong, a former policeman with a fierce reputation, "had taken Lilon." When Maiwot returned to Ambrym, he demanded the return of his wife, but Bong refused. Maiwot then demanded to "be reimbursed for the pigs and money he had paid for her." But still Bong refused (Ballard 1937). Ultimately, the DA ruled in favour of the woman's right to leave her marriage. This decision angered Paton, who pointed out that whereas the DA had in one case ruled in favour of the woman leaving because the payments had not consolidated the union, he had also ruled in favour of the woman leaving in a case where the exchange had been properly made.

Though the missionary claimed that the BRC was applying conflicting logic, these were "test cases" because the BRC had to evaluate whether to side with traditional customs and arranged marriage or with women who wanted to leave their marriages. The cases share some common elements. They show that marriage practices were malleable: attempts to determine traditional practices depended on who was asked. In addition, in both cases, while the women's agency featured centrally in the colonial rhetoric, both women were also connected with men who were reputed to be fierce. The cases also show that marriage practices were embedded in village relations and did not just pertain to the free will of the individuals involved or to a purely economic logic that the colonial authorities attributed to exchange. Exchange was the point of contention for other competing relationships and thus returning the "deposit pig" was not equivalent to returning a cash deposit on a milk bottle.

One of the last cases in the file refocused my attention on the imbalanced sex ratio, which had made the file intriguing to me in the first place. In his missive to BDA Adams, dated 29 January 1941, Paton began by noting "trouble in the village," for which he provided the following overarching explanation: "There are too many young men and too few young girls. The lads of Rohor [a village on the island of Ambrym] have decided that no Rohor girl can marry outside the village, unless they give a girl in exchange. They are too interrelated to marry Rohor girls" (Paton 1941). The matter at issue concerned the fact that an ideal marriage – from the viewpoint of Rohor men – between a Rohor man (named Rorang) and a young woman (named Mary) from a neighbouring village had been initiated and initially agreed to by the woman. But it was a tentative agreement in the view of the woman and her family because the Rohor man was from her "prohibited marriage line," according to Paton. Mary later admitted to the missionary that she wanted to marry another man, and that her father had consented.

Rorang flew into a rage, and, together with others from Rohor, went to the girl's village to stop the marriage. For a time, Rorang and Mary were together again (Mary's intentions are unclear), but she ultimately refused to marry him.

The story concludes, in a nutshell, with the missionary asking the DA for assistance in dealing with a man, a jilted suitor, who then raped or seduced (depending on who is asked) a woman in a neighbouring village who had agreed to marry him but then changed her mind. The trouble arose because "Rorang and others are trying to prevent the marriage under the pretext that girls are few and should marry into their own [Rohor] villages" (Adams 1941). The DA ultimately recommended to the resident commissioner that the man, Rorang, be detained for two months in Bushmen's Bay, allowing the woman sufficient time to marry the man who met with her family's approval.

Eventually, a colonial administrator drafted a memo for DAs so as to enable them to recognize "preliminary measures to be taken before marriage between natives is solemnized." These measures were as follows:

> 1 Consultation between a) Native Chief and b) Contracting parties and c) Responsible representatives of the families involved.
> 2 Report by the above three groups to the missionary to solemnize the marriage. A) There must be unity among all parties as to the legitimacy of the proposed marriage. B) The contracting parties must each make a categorical statement of the willingness to unite.
> 3 Public announcement of the marriage to be made not less than one week preceding the marriage. This notice [is] to be given when possible at a united gathering of the district or districts of which the parties are members, for example, at the weekly united service at the mission station.
> 4 The price arranged for the bride in money, pigs and/or other goods to be reported at the interview (point 2 above) together with terms of payment that is, whether already paid, or deferred, and if the latter, for what period. (BRC Office, n.d.)

The administrator issued these instructions to prevent violence and give marriages the best chance of success. The majority of the letters compiled in the file concern divorces that did not go smoothly, and cases of aggrieved parties who turned to the missionary or the DA for help. The paper trails in this file reveal the colonial desire to deal correctly with violence that had taken place within marriages, or to prevent such violence altogether. Often, violence or threats of violence erupted over how to handle the bride price in cases of "divorce." The correspondence in the file concerns conflicts over what to do when a marriage had broken down and the bride price had not been returned.

The cases also reveal that concerns regarding marriages were not private matters; they extended well beyond the conjugal couple. The colonial intervention of native courts made marriage and reproduction into public – which is to say colonial state – concerns. These legal frameworks were only possible because Ni-Vanuatu brought their concerns to the courts and because marriages were connected to exchange relations. You and I, living in the twenty-first century, can know about this particular aspect of history because of the public nature of the social form that created a new social space and required people to document, file, and index correspondence around it.

Free Will and Consent: Women Who Object

While colonial authorities' economistic accounts of Indigenous marriage agreements rendered women as objects, their concern with women's consent and free will feature prominently in the colonial and missionary correspondence in both the northern and southern regions. It was mainly women's (not men's)[11] consent that appeared to be a specific concern within colonial reports, typically in the form of rape accusations or unhappy marriages. For example, the age of consent figured prominently in colonial correspondence on whether to prosecute Native Code offences as "rape" or "seduction." In his correspondence seeking to determine due process regarding this issue, the BRC wrote the following lines to the French resident commissioner (FRC): "The question of 'age of consent' of female natives in the group is a very old and highly contentious one. I have heard it discussed on innumerable occasions by agents, planters, traders and missionaries (none of them agree)" (Joy 1934a). In the end, the BRC and FRC both agreed that it was "not practicable to determine a fixed age of sexual consent for young native females"; rather, it was preferable to defer to the DA's "discretion" and the medical evidence as to whether rape charges should be laid (Joy 1934b).

Colonial administrators raised the concern that early sexual activity (generally deemed non-consensual) for girls caused sterility and contributed to depopulation. The issue was that there was no clear age of consent in the Native Code. This meant that by order of the BRC, the BDAs could use their discretion and charge the concerned man with rape, irrespective of whether the girl had consented (Joy 1932). If the girl was deemed to be of the age of "sexual connexion" (how that would be defined was not specified) and no consent had been given, rape or indecent assault would be the charge. This would be proved by listening to the girl's claim but also by obtaining medical evidence and subjecting the girl to a medical exam by a doctor (this was only an option for those living in the vicinity of a hospital). Unsurprisingly, few of the cases documented in the colonial record proceeded that far.

Though the DAs' mandate was to uphold Indigenous marriage customs, they frequently indicated that they had made it known in the villages that they would support women who did not want to enter into arranged marriages. The following example of a DA's comments to the BRC is fairly typical: "I have done my best to impress upon the natives that women were not cattle and that it was time that girls should be consulted as to the choice of a husband, warning them that the Administration would always take the part of the woman forced into marriage against her will" (BDA 1940b). If she was unhappily married, a young woman could leave her husband and go to the mission (either Catholic or Protestant), where she could stay until she remarried. These situations were not uncommon, and they were always dealt with in this way according the above-cited DA (BDA 1940a). If the pigs were returned, it was stated that the woman had been given "her freedom" or that she was "liberated." The language and the social forms invoked in this context reflected an attempt to inculcate the possibilities that a woman would act on "her own free will" rather than be imbricated within kinship networks. In the words of the missionaries and colonial officials, if women were to choose for themselves in their domestic lives, this would be the modern way, and it would therefore be supported by the administration. The missionaries, in particular, were keen to support nuclear families and conjugal households.

I have been able to tell these stories by locating and reading letters in colonial files, but also because colonial forms framed what could be documented. The letters contained in these files pertain only to the dissolution of marriages that were deemed problematic. We learn of women who took action and left their husbands, opting to return to their natal homes or live at the mission, but we cannot know whether such women felt "free." We do not learn of the possibility that women remained in marriage and demonstrated what Wardlow calls "encompassed agency" (2006, 66), whereby women's skills in cultivating relationships contribute to social reproduction in such ways that their contributions are "encompassed" by social processes and normative social expectations over which they have little authority or control. Neither can we say for sure whether this is an example of what Wardlow terms "negative agency" (72), when women refuse to participate in social reproduction and reject gendered social expectations, a process she documents in her ethnographic fieldwork in Papua New Guinea, though at a much later period.

The documents do not allow us to know whether all women felt coerced by their natal families to enter into their marriages. We do not learn about women who appreciated their situation or who felt content within the web of their relationships because these situations were not brought to colonial courts. Attempting to understand the many dimensions of women's happiness or agency in marriage from a file about marriage and divorce would be like trying to write about marriage in North America or Europe solely on the basis of what happens in divorce courts. However, what divorce courts often do demonstrate, as in the case of the native courts in the New Hebrides, is that marriage is an economic relationship subject to public scrutiny. In the native courts, the colonial officials ascribed to themselves the authority to regulate these concerns that are at once public and economic. Many divorces ended up in native courts because Ni-Vanuatu wanted an external person to weigh in on a dispute that could not be resolved to their satisfaction in the village.

Feminist anthropologists have held that agency is shaped by the options available in particular social and historical contexts. This compels us to take seriously forms of women's agency that might not overtly challenge male authority (e.g., Mahmood 2001). In Vanuatu at this time, we see, on the one hand, agency documented by colonial authorities and missionaries who lauded themselves for helping women to leave their marriages and go to the mission station. On the other hand, the colonial regulations made visible those women who were running away from their marriages because they were documenting the problems associated with marriage-related exchange and divorce. On the grounds of promoting women's agency and preventing violence, British colonial officials and Presbyterian missionaries felt they had the right to intervene, and they did so through the creation of a public and economic sphere subject to state policy. The kind of care that colonial officials and missionaries showed for women's circumstances overlapped with their respective interests in promoting the expansion of colonial authority and Christianity, as it did in many places that were once part of European empires (e.g., Ahmed 1993; Abu-Lughod 2013; Hoodfar 2003).

Whereas colonial documentation does make these forms of agency visible, the fact that many situations were resolved outside of the courts leaves open the possibility that kin networks did not always compel women to stay in their marriages if they were unhappy. There are also other examples where marriage payments were eventually returned to the women's families. It is possible to imagine that such cases could have resulted from the women going to their relatives and negotiating the return of the bride price so that they could live with new partners or return to their natal homes. We do not read in the colonial record about the possibility of women exercising agency by negotiating with relatives for the husbands they wanted. Such forms of agency would not have needed colonial or missionary involvement and do not, therefore, appear in the colonial records. Limiting our analysis to the forms of agency that appear in the colonial archive, where agency is framed in terms of choice and autonomy, constrains our ability to theorize about women's lives, which is consequently confined to Western notions (Mahmood 2001). Yet, by following colonial documents as aspects of practices (of what tended to happen rather than what the discourse indicated should happen), it becomes apparent that Ni-Vanuatu men and women would not have engaged with the courts if they had no interest in them. Interventions took the form they did because of Indigenous social forms that entailed the participation of men and women in court proceedings. Ni-Vanuatu men came to the courts to demand the return of either their bride price or women when they failed to achieve these outcomes at the village level. Men could also come to the court as the "guardians" of their female relatives. Women participated by coming to the court and waiting there for their cases to come up for discussion. In the cases discussed in this chapter, the administration sided with women associated with fierce men. Generally speaking, the administration's attempts to negotiate with chiefs to reduce the bride price failed. In the bride price regulations and in native courts, we see how the colonial desire to protect women was implicated in colonial attempts to raise the birth rate. While ostensibly emphasizing women's agency, the colonial state's interventions also provided new fora in which to scrutinize morality. Furthermore, through the participation of the DA, missionary, and doctor, along with women's relatives in the courts, the number of men who could weigh in on Ni-Vanuatu women's lives increased. The cases that I have considered here appear to show that women had the opportunity to seek redress by availing themselves of new, modern dispute-resolution mechanisms, but in reality these mechanisms led to an expansion in the number of men who could control women's lives.

To researchers of that time, women's use of traditional means of fertility control conveyed an apathy or inability to adapt to modernity. However, caring for large numbers of sick and dying people would have been devastating and terror-inducing for these women. In this chapter, I have attempted to rethink how women tried to control their lives with the means available to them to show that women were not apathetic about their futures, reproductive and otherwise. I focused on colonial and missionary paper trails and associated social forms aimed at improving marriages (and thus fertility) to show how these forms rendered some kinds of women's agency visible and others' invisible. What colonial and scientific representations of women's agency have in common is that they render non-Western forms of agency invisible.

Conclusion

In this chapter, through highlighting the circulation of a population figure, I have provided some context for the women in Speiser's photo, which he took thinking he was documenting a group of people on the verge of extinction. The letters, reports, memoranda, files, tables, indexes, boxes, and archives I consulted comprised some of the accounting and quantification tools the colonial bureaucracy used to expand colonial governance at this time of frequent and severe infectious disease epidemics and low fertility. The imbalanced sex ratio circulated along with these documents, with the ultimate aim of implementing policies that, in combination with meetings and discussions, were intended to increase fertility through the regulation of bride price and to enable court operations for resolving marriage disputes. The documents demand of us a vigilant counter-reading, lest we normalize these otherwise dispassionate conversations between white men about the fertility of Pacific Islander marriages and the fate of Ni-Vanuatu women, such as the two in Speiser's photo. The documents are implicated in the creation of structural and racist inequalities. The role of figures in colonial bureaucracy are important to understand for how they render the world in particular ways, framing agency in ways that legitimate colonial and missionary presence, while eclipsing certain versions of the world. The imbalanced sex-ratio figure distinguished between women and men, and together with the reproductive concerns of the time, rendered men and women visible as targets of colonial policies. These policies that aimed at controlling reproduction and introducing native courts mapped onto the desire some Ni-Vanuatu had for engaging dispute-resolution mechanisms in the context of marriages and related exchanges that did not go as planned. The sex-ratio figure was thus embroiled in colonial attempts to change Islanders' relationships to objects, animals, and people.

When reproduction was scaled up through quantification to a population figure like the sex ratio, reproduction was constituted as a public issue associated with publics and populations, and entangled with making moral matters of marriage, sexuality, and the right relations of exchange subject to new forms of public and state concern. The colonial authorities introduced mobile courts that propelled the creation of a polity larger than any individual island or village. These attempts to control reproduction, sexuality, and violence did not occur on a blank slate; the forms they took followed the contours of Indigenous marriage practices and moral conventions. Economic relationships and concerns for reproduction overlapped in the concerns of colonial authorities and Ni-Vanuatu, which enabled the practices relating to bride price to be brought to and adjudicated within an expanded public domain. This chapter has shown how colonialism constituted another layer of intervention in women's lives, one that dovetailed with Indigenous men's power, while opening up some new paths for women.

That the sex ratio was measured and circulated as though it should be brought back into balance through state policies and interventions also indicates a historically specific way of thinking about a population's reproduction. This is part of the scientific changes from this period that Murphy (2017) argues marked the reproduction of a population as an object of experimentation and site of control. Murphy further argues that this scientific shift laid the ground for demographic and economic knowledge to connect in the post-war emergence of GDP as an integral benchmark for the co-organization of capitalism, a form of state governance through reproductive control. In the pre-war era, people in Vanuatu linked the reproduction of the population to economic concerns that were tied to the expansion of capitalist forms of exchange.

It was in the file titled "Vital Statistics Land etc." that I first located the importance of the sex-ratio figure in relation to colonial attempts to control reproduction through the regulation of marriage practices. Connecting vital statistics, marriage, and land was indeed central to the colonial project's aspiration of establishing control. Land and marriage also underpinned Ni-Vanuatu concern with the right kind of exchanges and relationships of reproduction, which, as the following chapters will show, exceeded the moral figures of capitalist and imperial social forms.

Chapter Two

"The Nurses Looked Out for Us!": Hospital Births, Relational Infrastructures, and Public Concerns, 1950–1970

> The capacity of nurses to care about strangers in ways that many of us reserve for kin leads us to champion their heroism.
>
> Alice Street (2016, 333)

When people gave directions to Madeleine's[1] house in Pango Village, they said, "It's the one with the beautiful flowers." Madeleine was in her senior years when I met her, and she did not leave her yard that often. Her extraordinary rose garden reflected the time she now spent at home, yet the constant presence of her children, grandchildren, and great-grandchildren flitting in and out of her spotless, turquoise-blue house, dispelled any notion that her days at home were lonely ones. In the early 1950s, when she was in her late teens, Madeleine trained as a nurse at the Paton Memorial Hospital (PMH) in Port Vila. In addition to raising six children, she worked as a nurse and midwife at the PMH or a small clinic in Pango for much of her life.

To recognize her lifetime of achievement and service, Madeleine was awarded a national medal of merit by the president of Vanuatu. The ceremony figures prominently in one Pango musician's music video for a song about the village. In the video, Madeleine is shown smiling radiantly and wearing an *aelan dres* as she accepts the medal. Originally introduced by Presbyterian missionaries (Bolton 2003a), the *aelan dres* is the national dress of Vanuatu and a sartorial marker of feminine respectability (Cummings 2005), and it is increasingly worn mainly by women of Madeleine's age. It is no accident that the nation of Vanuatu and her village lauded her biomedical skills in conjunction with her embodiment of ideal feminine Christian virtues. She and the women like her who worked as nurses and midwives in 1950s and 1960s are the moral figures of this chapter. They are remembered fondly in the oral histories of senior women in Pango today, who told me in different ways that "the nurses looked out for us!"[2]

Over the course of Madeleine's lifetime, Pango Village, and Vanuatu as a whole, has experienced tremendous social and demographic change. The proliferation of Christian, colonial, and capitalist social forms alongside increased urbanization has brought profound changes to ancestral traditions and village social organization. Within this broader context, Madeleine and her peers have participated in the medicalization of childbirth, as the most common place for giving birth in Pango Village has moved from thatch house in the village to Presbyterian mission hospital, to British colonial hospital, and again to the Vanuatu government hospital. There has been a similar transition in the figure of the ideal caregiver, from the village expert taught by older kinswomen to the professionally trained Ni-Vanuatu nurse. Demographically, these changes have been part of the shift from population decline due to high mortality and low fertility to rapid population increase due to longer life expectancy and increased fertility. Nurses like Madeleine played a key role in the expansion of medical infrastructure through which childbirth was made a concern beyond kin and village and relocated within Christian and then colonial biomedical institutions. Ni-Vanuatu nurse-midwives were trained to be, and are remembered as, moral figures for the particular care they provided: they were able to combine Christian values and Indigenous practices with medical knowledge, and their social networks, labour, and knowledge were essential relational infrastructures that could compensate for the insufficiency of medical technologies.

In some ways, the history told in this chapter is a similar narrative of the medicalization of birth that has been important part of the critical and feminist historiography of women's health in many parts of the world. In certain places, this process has been critiqued for the overbearing presence of technology, which can have negative effects on women's experience of birth, leading to what Davis-Floyd (2004) has called "technocratic birth." The medicalization of birth has been convincingly shown to have had negative effects on women's health in places where technology is not consistently available (e.g., McPherson 2007). In yet other cases, such as Van Hollen's (2003, 213) ethnography of childbirth in the Indian state of Tamil Naidu, poor women resent the lack of biomedical technologies available to them and experience this as a form of discrimination. The medicalization of birth in Puerto Rico, Córdova (2017) argues, was associated with the colonial development of a particular kind of industrialization and governmentality; Cosminsky (2016) traces a similar trajectory in Guatemala, where childbirth changed with the expansion of the plantation economy. As I show in this chapter, attempts to medicalize birth in Vanuatu made reproduction a public matter – a process or "problem" that could become a humanitarian cause taken up by doctors, rather than an event that took place in the village – and occurred in tandem with the expansion of cash economies. I also show, however, that the shift in birth practices depended on Ni-Vanuatu care and knowledge economies that exceeded commodified social relations.

To look at how birth caregivers were made visible to and by the colonial project is to focus on a collection of biopolitical concerns. This was a colonial project concerned with improving the population's mortality rates and overall health during the post-war era, and one of its goals was shifting the normative place of birth and birthing knowledge. It was a project associated with training and paying educated workers and finding capital to build and maintain infrastructure. With the medicalization of birth would come the increased surveillance and documentation of pregnancy, mothering, and nutrition. In this focus on birth, the colonial concern for a population's welfare met medical knowledge, which in turn met surveillance and economic concerns. These were global forms of post-war biopolitics, where along with a concern for medicalized childbirth in international health, global research infrastructures and international institutions simultaneously made population size and growth visible so as to control the size and quality of populations. In the Global South, as Murphy (2017) cogently demonstrates, these infrastructures were associated with the medicalization of reproduction in the form of dramatic population control measures. This occurred in tandem with the growth in the visibility of "the economy" and the importance of GDP-based success metrics for development (Murphy 2017).[3] In post-war Vanuatu, the medicalization of reproduction and colonial biopolitics meant making unpaid care labour and non-monetary exchange visible. For the recruitment of nurse-midwives, their medical education, and the payment of staff depended on Indigenous networks and forms of sociality, and care economies in particular.

The fact that birth was brought under the purview of biomedicine is also part of the global history of the expansion of biomedicine and the role of the public hospital. Drawing on different aspects of Foucault's analysis of modern institutions, medical anthropologists have written about hospitals as biopolitical sites where bodies are made visible to the clinical gaze, where modern powers of surveillance and control are tested and perfected, and where new subject-making processes are set in motion. In European colonies, public hospitals have been sites where the colonial presence is legitimated and enabled (Anderson

2009), and where modernization and development are promoted (Street 2014), particularly in the post-war period. Ethnographies of hospitals and clinics in the Global South (e.g., Hull 2012; Livingston 2012) show that, given the lack of technologies, combined with the ingenuity of professionals, patients, and their families, these institutional spaces do not become the sites of control or surveillance. Hospitals and clinics can often become places of social suffering and neglect (Biehl 2005). In Papua New Guinea, according to Street (2014), postcolonial hospitals are places where the postcolonial state's efficacy and development efforts become tangible. Such expectations of state protection were not met, however: the hospital in Street's ethnography was a space in which patients and their families did not feel seen by the medical workers and the state more broadly. In addition to these aspects of public hospitals, I show how in Vanuatu, mission and colonial hospitals were places where new moral figures were trained. Nurses were visible as colonial and Christian subjects in the 1960s, but because they were seen as moral subjects for the care they provided, they remained visible decades later as figures through which older women could critique the feelings of neglect they experienced at the government hospital.

Missionaries' and colonial officials' attention to birthing women and women's reproductive bodies is a common feature of the social hygiene and modernization efforts mounted by European colonial regimes – modernization efforts that, in practice, were accompanied by attempts to inculcate Western notions of the body, motherhood, sexuality (e.g., Boddy 2007; Ram and Jolly 1998; Guha 2018; Coghe 2022). As in many places, biomedical practices did not simply travel unchanged to Melanesian contexts; rather, they were "remade with the personnel, materials and expertise available and in relation to the particular needs of a fragile state and uncertain plantation economy" (Street 2019, 311). Thus, Street writes, medical institutions might be seen as "frontiers, where new kinds of biomedical knowledge and institutional forms were produced and negotiated in relationships between colonial missionaries, health workers, administrators and citizens" (311). The negotiated, contested, and incomplete nature of this medicalization of birth and reproduction notwithstanding, the expansion of biomedicine is part of the historical processes that created and reproduced racialized inequalities and structural violence. The nurses' care was embedded in all of these processes in Vanuatu: they were part of the negotiation of multiple knowledge, economic, and institutional forms.

Acquiring, using, and advocating for costly biomedical technologies and infrastructure were part of how reproduction became a public concern. Less visible to colonial projects than material infrastructure and medical technology, but essential to women who were giving birth, was the importance of care. Care is far more than a set of emotions or actions located in an individual's capacity, as many medical anthropologists have shown. Medical care is what McKay (2018) calls "relational work"; it takes multiple forms and meanings according to the social networks, bureaucratic and NGO concerns, as well as political, classed, and gendered subjectivities. Care, medical and otherwise, has also been shown to be embedded in "ecologies of support" (Duclos and Criado 2020) that are part of "socio-material modes of togetherness" and "precarious connections and solidarities" (161). Care relations are also where ethical and moral subjectivities are formed, in both deliberate and indeterminate ways (Scherz 2018). The nurses' place in the medicalization of birth in Vanuatu shows the relevance of these ways of thinking about care. Nurses are at once locally meaningful subjects and part of relational infrastructures. This is what made them moral figures in both biomedical and Ni-Vanuatu care economies.

Particularly in colonial and postcolonial contexts, care is never innocent because of its implication in multilayered structures of inequality that are racist, sexist, and connected to the expansion of capitalism (Murphy 2015b). Medical care provided by humanitarian organizations is often offered as a seemingly apolitical means of saving lives (Ticktin 2006). When provided by the settler state, medical care for Indigenous people – when it is offered at all – tends to be "anonymous" and focus on mere biological survival to the detriment of relational forms of nurturance (Stevenson 2014). State care (medical and otherwise) of Indigenous people thus participates in re-enacting racist forms of structural violence. Ni-Vanuatu women's generally positive memories of the colonial mission hospital might appear to occlude political aspects of care. However, as I will show here, their positive memories of care are at once a nostalgic view of the past and a political critique of a present in which they perceive a diminishment of care economies and relational infrastructures needed for rituals and daily survival as wage labour increases. The women's memories of birth at the hospital, which emphasize the Ni-Vanuatu nurses, teach us that, while hospital technology is necessary, the networks of care as relational infrastructures that surround that technology are also crucial. Their oral histories show what Michael Lambek (2015) has argued – namely, that memories can be collective forms of moral practice, which here coalesce around a moral figure, the nurse. Their memories show contemporary moral ideal values of Ni-Vanuatu modernity: Ni-Vanuatu nurses were skilled at combining knowledge systems on their own terms.

I begin with how relational infrastructures and care economies of Ni-Vanuatu birthing care were made visible in written scientific and colonial accounts in the first half of the twentieth century through researchers' interest in Indigenous birth attendants' tools, techniques, and methods of "payment." This provides the scientific and socio-cultural context of three concerns the British authorities had in the 1960s concerning the medical administration of birth: how to train nurses, how to deal with the substandard maternity ward at the PMH, and how to handle the transfer of rural dressers from the Presbyterian mission to the New Hebrides British Service (NHBS).[4] This is to show the biopolitical terrain on which reproduction and birth were medicalized during this period. This involved population-level concerns of expanding and standardizing medical knowledge, rendering non-monetized exchange networks available for the expansion of medical labour, and advocating for an adequate health infrastructure in the face of resources always professed to be scarce. The biopolitical processes around the medicalization of reproduction produced moral subjectivities that individual nurses would embody.

In the second part of the chapter, I discuss Pango women's oral histories of the nurses' training and the women's birthing experiences at the PMH during the 1960s. The moral figure of the nurse and the care she provided in the 1960s needs to be placed in the broader context of changes to the care networks – relational infrastructures – in the village from the 1960s until 2010, as these changes inform the women's claims about the importance of care and care economies. Pango women's memories offer insights into how biopolitical interventions in the 1960s were transformed into moral resources for critiquing their postcolonial context.

The chapter tacks between two very different scales – those of the archipelago-wide polity and the peri-urban village of Pango. This might be jarring for those who know of the diversity of cultural practices and localized forms of social organization in Vanuatu. Indeed, people's lives in rural, peri-urban, and urban contexts vary quite dramatically, and Pango women's histories tell a story situated specifically in their peri-urban experience. I hope to show that because of the institutional spaces associated with polities that maintained and staffed the hospital or nurse training program, people's lives during this period were lived not just at the village scale, but also with the reality of a polity beyond the village. In delivering and receiving reproductive medical care, Ni-Vanuatu in the increasingly urban area around Port Vila came into contact with other Ni-Vanuatu, both as caregivers and as fellow patients. Their eventual expectations of care from the postcolonial government began from their experience of such institutions. Moving between these scales of social organization in the delivery of medical care was part of making reproduction public.

Birth and Care for Birthing Women as Scientific and Colonial Concern

Indigenous practices surrounding prenatal, birth, and postnatal care were considered profoundly primitive by missionaries and medical doctors in the early twentieth century (Jolly 2002, 148–51). In Felix Speiser's anthropological research, he shows a scientific interest in the women who assisted birthing women. From 1910 to 1912, as he documented the diversity of the many islands that now make up Vanuatu, Speiser ([1923] 1996, 254–7) discussed the diversity of midwifery practices. In that work, Speiser opens the section on birth by speaking against prevailing Euro-American assumptions about how birth took place in non-Western places – that it was easy for women. It was not (254). He goes on to describe typical birth locations, usually women's huts (as was the case on Santo, Malo, Malakula, Ambrym, southern Pentecost, and the Banks Islands) or specially constructed huts in "the bush" (as on Ambae, northern Pentecost, and Maevo), or sometimes in the bush itself (as on Tanna and Erromango). He remarks on the squatting birthing positions on different islands and takes note of some common tools: bamboo knives and ropes. For example, on the Banks Islands, Santo, and Ambae, a bamboo knife was used to cut the umbilical cord (254). On Ambrym and Pentecost, women used ropes to support the labouring woman:

> They tie a stout rope to the ridgepole of the house with its end hanging loose. The woman holds the rope with her hands raised above her head. On the onset of the second pains one of the midwives holds the woman firmly round the chest from behind and, squatting herself, takes the mother between the thighs and squeezes her abdomen hard from the sides. When the child appears, it is allowed to slip down of its own accord onto a mat or the like, care being taken simply that it does not fall from too great a height. (254)

While women gave birth in the company of other women, "the men are never present; should they come in the vicinity they are chased away by the other women" (254).

In addition to documenting their techniques and technologies, Speiser took particular note of the exchange relations surrounding the expertise of the woman attending births. Speiser described the importance of the "payment" of midwives by the newly delivered woman as follows: "the midwives, like the other women who give assistance, must be remunerated, especially those who look after the woman afterwards, relieve her of her work or help in the house by doing the cooking" (255). Writing of the island of Pentecost, Speiser says that "a generous payment must be made to the two midwives, the woman who cooks for the mother, and the two other women who hang the afterbirth up in the wood (in European money the total sum is in the region of £2, which is a very considerable amount)" (255).

By the 1950s, there were Pacific Islander physicians delivering medical care in hospitals across the New Hebrides. Officially, they worked under the supervision of a French or anglophone physician, but in practice they travelled alone to villages to deliver curative and preventive medical care (Widmer 2010, 2013). While working in this capacity, the Fijian assistant physician Peni Tuidraki (1957) devoted an entire article to the "New Hebrides Midwife" in the Fiji Medical School publication the *Native Medical Practitioner.* He focused on "midwifery" (notable that he would use the biomedical term to elevate their traditional practice) on the island of Santo, indicating that both Christian and traditional communities used the same methods. Women gave birth in a "confinement house" assisted by "an old woman" who does not do internal examinations and who "only keeps her eyes to the advancing part of the foetus from the vulva" (Tuidraki 1957, 489). Only after the placenta is out will the woman cut the cord with a bamboo knife and tie it with a coconut fibre. Abdominal massage is used to expel a retained placenta. The midwife takes care of the baby while two of her assistants care for the mother. All the women in the village must "give a present to the newly born baby a day after birth" (489). Tuidraki also took interest in how midwives were compensated, noting that "in accordance with native customs here the midwife is given some pigs, yams, etc. as payment for her service" (489).

In Pacific Island contexts like Vanuatu, as the sizeable literature on exchange shows, giving and receiving objects like yams, woven pandanus mats,[5] pigs, or bundles of cloth serves to create and reproduce socially significant relationships and moral personhood. Amounts of cash could be part of exchanges in combination with objects. Life cycle events are particularly important moments for exchange ceremonies, like those surrounding birth and marriage and death (e.g., Hess 2009; Bolton 2003b; Rio 2007). At these times, the exchanges reproduce relations, including those pertaining to inheritance and descent, between families and groups, as well as produce the identity of the person or people at the centre of the ritual. There are variations in the kinds of ceremonies, objects, and amounts exchanged within Vanuatu, and these also change over time. Because of the nature of these exchanges, "payment" – a term associated with commodified exchanges in which ongoing relationships are not required for giving and receiving – is not a fulsome descriptor for exchange in these contexts. Indeed, researchers recognized this during the colonial period and would indicate their awareness with quotation marks around the term. That the colonial researchers were interested in "payments" surrounding exchange, and then in turn considered using them in colonial policy, is what interests me here.

In the years between Speiser's and Tuidraki's research on birth attendants some aspects of how biomedicine was delivered had changed in Vanuatu. Since the beginning of the Presbyterian and Melanesian (Anglican) missions in Vanuatu in the mid-1800s, missionaries organized medical activities through small dispensaries, and eventually clinics and hospitals.[6] After the establishment of the Condominium in 1906, the NHBS increasingly funded the English-speaking medical service operated by the missions, while the French service ran their own clinics and hospitals, and the Condominium's health service paid the salaries of Pacific Islander assistant physicians. (An overview of the Condominium's biomedical services in the mid-1950s can be found in appendix 2.) Gradually the British authorities assumed control of the administration of the mission-operated health services, but the Presbyterian and Anglican missions remained closely involved in the day-to-day operations and staffing of the hospitals.

By the 1960s, the PMH had become the largest Presbyterian (and, indeed, the main anglophone) hospital in Vanuatu after a volcano destroyed the Presbyterian mission's hospital on Ambrym in 1913. The PMH was built on Iririki Island, in Port Vila's harbour, in 1910 and opened in 1911. Initially funded by the NHBS and the John G. Paton Mission Fund Committee, by the 1960s the NHBS had assumed nearly all the financial responsibility for the hospital. The BRC's residence was a short walk up the hill from the doctor's house. The PMH was reached by a launch from the customs wharf, where today large cruise ships dock.[7] There was also a house for the medical superintendent and his family. Anglo-European and Ni-Vanuatu nurses were housed in separate accommodations in the hospital compound from the hospital's founding (Stevens 2017, 605). Male dressers were housed elsewhere on the island. Today, the site of the hospital is occupied by posh restaurants and pools as well as the best-equipped children's playground in the area, accessible primarily to people at the surrounding luxury resorts. From 1935 to 1946, Dr. Alexander Frater served as the medical superintendent. Dr. Knox Jameson replaced him until Dr. Freeman took on the role in 1962. By the time Dr. Freeman arrived at the hospital, patients were coming from the villages surrounding Port Vila, Mele, Pango, Erakor, and Eratap, as well as islands as far north as Malakula and as far south as Erromango and Tanna (Freeman 2006).

From 1936, New Zealand and Australian nursing sisters of the Melanesian mission (Anglican) ran another important mission institution, the Godden Memorial Hospital (GMH) at Lolowai, on the island of Ambae. In the late 1940s, an assistant medical practitioner from the region began working there (Southern Cross 1950), and the NHBS paid half of his salary. Treating Hansen's disease was an important aspect of the mission's work at the nearby St. Barnabas Leprosarium. Dr. Bruce Mackereth, who arrived with his wife, Catherine, and their children in 1962, was the first doctor to be permanently employed by the mission. Around 1964, Anglican sister Betty Pyatt[8] described the GMH as follows:

> The bed count at present stands at, 14 Maternity, 23 Tuberculosis and a "bunk house" for 5 ambulatory Tuberculosis men, and 20 general beds. There should be 24 general beds but 4 of them are taken up with the orphan babes. Just completed are: ablutions blocks, laundry, hospital office/duty room and Pharmacy. Also … the Mothercraft house was given to us by N.Z. Corso. Next in the plan are the theatre, sterilizing room, outpatients and X-ray with Laboratory. (Quoted in McMenamin 2009, 318)

The Mothercraft House at the GMH was the place where orphaned newborns would be cared for and skills taught to new mothers. Mission publications articulated the reasons for this aspect of their work as follows: "The mortality among infants is appalling. Efforts to stay this include the regular instruction of women at all district dispensaries, where white nurses help and advise native mothers in pre-natal days and, where they are allowed to do so, at the time of confinement." Orphaned infants were cared for and kept until the age of about fourteen months, when they were "strong enough to go to their families" (Diocese of Melanesia 1943).

In the way that birth was documented before and alongside the development of biomedical institutions, then, researchers highlighted the tools, diversity of practices, and expertise. Birth attendants were made visible for their knowledge as well for a glimpse of their entanglement in Indigenous exchange and care economies. The foci of the researchers, on tools, techniques, and "payments" at birth, signal similar concerns in the biopolitical terrain for improving conditions for medically attended births, part of the wider effort to develop a population that I discuss in the upcoming sections. The tools, techniques, and payments also figured in Pango women's oral histories of birth as indications of care, as I discuss in the final section of this chapter.

Standardizing Nursing Knowledge: Training Moral Figures

In the 1950s, there were three English-speaking nursing schools in Vanuatu, one run by Anglicans in the northern region at the GMH on Ambae (formerly Aoba) and another run by Presbyterians in the central region at the PMH on Iririki in Port Vila Harbour. There was an additional small centre at Ndui Ndui on Ambae run by Seventh Day Adventists. Sister Pyatt (Anglican) at the GMH and Sister Edgar (Presbyterian) at the PMH played central roles. Ni-Vanuatu women had long worked in the mission hospitals on Ambae and Iririki in the areas of food preparation and cleaning, as well as some caregiving, before formal trainings were gradually put in place in the 1950s. In the 1960s, as the British planned to take over the missions' health services, there was a concern about how to standardize the professional training for Indigenous nurses and nurse-midwives across the two larger schools at the PMH and GMH.

Beginning in 1951, Sister Betty Pyatt developed a two-year course at the GMH in which women from the northern islands of Vanuatu were taught to be nurses. By the 1960s, the program had grown more robust and the three- to four-year training consisted mainly of on-the-job instruction with five lectures a week. Pyatt (1963) writes, "the subjects they learn are: hygiene, infant welfare, junior maternity, anatomy, physiology, and medicine (integrated), practical procedures (nursing techniques), minor surgery (suturing, extracting teeth, opening abscesses), village work and the care of dispensaries, senior maternity (including all complications)." The new nursing program started with three to four pupils per class, while in 1963 they aimed to have five per class; by 1966, nine nurse trainees began the training, with seven starting the following year (Mackereth 1966). Men were also trained alongside the nurses as "dressers," but in their curriculum "senior maternity" was replaced by a course in "public health" (Mackereth 1963). It was difficult to recruit qualified students, as levels of primary education were low. The goal was to admit students who had completed six years of primary school. Pyatt recognized that the nurses needed to be "doctor substitutes" (her term) and ultimately "responsible for diagnosing and treating, ante natal clinics and deliveries, baby clinics, health talks in schools, public health around villages" (Pyatt, n.d.). In her annual report, Pyatt proclaimed, "they need to see all the Maternity Work they can when they have to nurse, on completion of their training, on such isolated islands as the Banks and Torres. Very few graduate from here without having confined 30–40 mothers and witnessed and assisted with twice as many more" (Pyatt 1960). The graduates of this nursing school would work in the hospitals in Lolowai or Santo or operate basic clinics, primarily in the Northern District.

A similar two- to three-year nurse-training program was in place at the PMH after having been formalized in 1947 by Sister Elspeth Edgar. Theoretical knowledge and practical skills were not program's the only goals. Dr. Ted Freeman, the medical superintendent of the PMH, recalled in his 2006 memoir that "a strict Presbyterian code of conduct was enforced on Iririki." This meant that "alcohol was forbidden and attendance at church was expected … The Bible was seen as the Word of God, and the concepts of love, peace, patience, stoicism, discipline, tolerance, and brotherly and sisterly love were promoted" (Freeman 2006, 47). He further remembers that,

> as the hospital drew nurse trainees from many different islands, often with fairly widely diverse sub-cultures, there was always the possibility of friction but, to my knowledge, this did not occur. I think this was mainly thanks to the influence of Sister Edgar, who maintained very high standards herself and expected all others to do the same. On the outer islands, Marge Heard and Sue Steel (Ramm) at Lenakel Hospital on Tanna, Alison Todd on Paama, Barbara Cunningham, Pat Hewitt and Val and Jack Ridler at Vaemali Hospital on Epi, were all very responsible people and they encouraged a similar peaceful and harmonious attitude amongst their staff.
>
> Codes of conduct extended to relations between men and women. During morning prayers at PMH, the nurses, kitchen girls and seamstresses would sit on one side of the nurse training hall and the male nurses, boat men and outside men would sit on the other. There was no fraternization. This separation of the sexes mirrored the practice in the village churches, where the men would sit on the left and the women on the right of the aisle.
>
> Strenuous steps were taken to prevent unwanted pregnancies amongst the New Hebridean nursing staff. On Iririki, the main hospital buildings separated the male and female dormitories. The male staff quarters were on the southern part of the island and the female staff quarters on the northern part, directly below the medical superintendent's residence. However, even with the geographic separation, it was difficult to control the sexual drive of young men and women and, while rare, some pregnancies did

Figure 2.1. Graduating nurses
Source: Photo by Dr. Ted Freeman. Australian National University Archives (ANUA), box 602, box 2, ca. 1963–70.

> occur at the hospital. In these cases, the girls were sent home to their villages. A pregnancy suggested that supervision of the nurses might have been inadequate and had the potential to make families reluctant to send their daughters to PMH, leaving the hospital unable to provide trained staff for Vila or for the dresser stations around the island. (48)

I quote this passage at length to demonstrate the significance of the "codes of conduct" for the nurse trainees, particularly with respect to interpersonal relations between women from other islands and between women and men. From the missionaries' perspective, training Ni-Vanuatu nurses in the technical skills of nursing was inseparable from the moral dimensions of their education (see figures 2.1 and 2.2).

Standardizing Training across Vanuatu

In April 1963, Sister Pyatt and Dr. Freeman met in Port Vila with representatives from the other smaller medical missions, those of the Church of Christ and Seventh Day Adventists, to discuss how they might coordinate the seven medical services – one each administered by the four

Figure 2.2. PMH nurses line up on the wharf
Source: Photo by Dr. Ted Freeman. ANUA, box 602, box 2, ca. 1963–70.

religious missions and by the British, the French, and the Condominium – across the islands. The coordinating body that emerged out of these discussions became known as the British Medical Advisory Committee. Standardizing nursing training was seen as a priority when it came to meeting the needs of a growing population. A training program in the New Hebrides was appealing, those present believed, because young women would face "moral dangers" if they went to train in Honiara in the Solomon Islands. At the next meeting, in 1964, they decided that Sisters Heard and Pyatt would work together to combine their curricula in light of what was being done in Honiara and the recommendations of the World Health Organization (WHO). They presented the following syllabus to train young women to be community and clinical nurses:

Training:

Yard and routine hospital duties in general
Maternity work
Outpatient work
Village visiting
Lectures and demonstrations

Subjects:

Ethics of Nursing and Hospital
Practical Procedures
First Aid
Scripture
Infant Welfare
Simple Surgery
Midwifery
Nutrition and diets
Diseases with Anatomy and Physiology combined
Or Diseases Anatomy and Physiology (as separate subjects)
Community and personal hygiene
Dispensaries (Heard and Pyatt 1966)

The ethical standards were to be worked out for each hospital. The expanded practical procedures and general nursing went as follows:

Beds and bedmaking:
Stripping a bed
Making a closed bed
Operation bed
Fracture bed
Changing sheets, etc. with patient in bed
Use and care of linen
Use and care of rubber goods

Cleaning, sterilizing and storing:

Bowls, bedpans, etc.
Instruments, glassware (Heard and Pyatt 1966)

For midwifery, the curriculum included such topics as early signs of pregnancy, infectious diseases during pregnancy, presentations of the fetus, the first stage of labour, normal labour, and possible complications.

At their 1966 meeting, in the sweltering heat of late February, the British Medical Advisory Committee[9] discussed this curriculum and agreed that a standardized exam should be developed. Those present suggested that WHO input was also important for this certification process. To that end, in 1969, a WHO consultant, Olive Manning, came to make recommendations on how to further improve the nursing curriculum according to international standards. In that proposed curriculum, students would first learn theory, but ideally in ways that tied it to practical activities that would allow them "to meet the health needs of the New Hebrides." For example, they would learn about "ovulation and embryo growth" in tandem with "contraceptive types" in the first year. In the second and third years, there was more emphasis on practical experience. Midwifery students would conduct twenty antenatal exams and attend twenty normal births. The young women being trained to work in the community would be taught survey techniques to gather population data, as well as best practices for keeping records and reporting their work. Overall, the curriculum emphasized how to use medical technologies and organize labour in a clinic, and how to document the work. The Medical Advisory Committee accepted Manning's recommendations, noting their ambitious scope.

Over the course of seven years, through discussions at their annual meetings in Port Vila, the British officials and English-speaking medical missionaries made plans to improve the curriculum in order to meet the public imperative of standardizing nursing knowledge and expanding the scope of nursing. Bringing maternal and child health under the explicit purview of the NHBS from the different missions was part of the biomedical effort to standardize knowledge and training. This knowledge needed to be legitimated by a system of formal state certification and vetted by an international organization. The process was accompanied by the inculcation of Christian values and comportment in Ni-Vanuatu women. Indeed, it was thought that it was only under these protected conditions that Ni-Vanuatu families would allow young women to train away from their home villages.

Compensating Birth Care in a Colonial System

In addition to new attempts at standardizing training, bringing birth more comprehensively under the purview of biomedicine was a labour force issue for the NHBS. By the mid-1960s, there was a network of "dressers" (male nurses) in rural areas with first aid and elementary nursing training who ran small dispensaries with limited medical supplies. Dr. Knox Jameson had begun these rural trainings in 1948 (Freeman 2006, 19). In 1966, the NHBS prepared to take over the dressers' network, which at that time was a largely informal organization of individuals working under very different working conditions in different communities. Dispensaries were not always open the same number of hours, staff had different levels of training and ability, and some had access to garden food while others did not. Moreover, while some were compensated in part by the communities, others were not. The assistant medical practitioners, like Peni Tuidraki, were paid in cash by the Condominium. A central question the NHBS grappled with at this time was standardizing how and how much to compensate staff working in the small dispensaries. About nurses, after service-wide investigation, the chief Condominium medical officer, Dr. W.H. (Bill) Rees reported that

> nurses have been employed recently as nurse-midwives, with ill-defined areas; and no salary has, in general, been arranged. It would appear that it is expected they will receive payment from the patient. The general pattern in the area is that native midwives are paid or given gifts. A large number of women sit with the mother. One is accepted as the senior midwife and all receive a gift from the husband (e.g. a yam or a fowl). Money payments are made, probably more than before. A PMH midwife received 6 pounds recently for spending a week or two with an expectant mother and attending the delivery. Of this, 5 pounds were given to the women gathered with her. (Rees 1966b)

While the overall recommendation for paying the dressers and nurses was to stop village-level compensation and to centralize administration in the NHBS, the recommendations for women who cared for women during birth was different:

> Where a nurse is employed in the same way as a dresser in a rural dispensary, the salary would be [the] same as for a Dresser. However, married nurse/Midwives often would not be prepared to undertake dispensary responsibilities, more particularly touring ones. Their abilities as midwives tend to be small, although they are better than nothing. Also, a married (or single) nurse may live within the area of a rural dresser. The employment of these people purely as midwives is not a practical proposition. As the people of all areas are accustomed to paying (or giving substantial gifts to) midwives, I recommend that a fixed fee of 3 pounds a delivery be adopted, and this be paid directly to the nurse/midwife by the patient's family. Midwifery drugs and equipment should be supplied by the Administration. (Rees 1966b)

In the end, the chief medical officer recommended the following:

> A trained and certified midwife or a clinic equipped to handle midwifery cases may make a charge of 3 pounds for a midwifery care at the clinic or village. This amount is payable to the worker concerned and is independent of payments made to others present, or to payments made [to her family] for long absences from her village. (Rees 1966a)

By 1967, nurses earned twenty-four dollars per month plus rations at the PMH, and sixteen dollars per month at the GMH (Medical Advisory Committee 1967).

As anthropologist Speiser and assistant physician Peni Tuidraki noted earlier in the twentieth century, it was common for new mothers to give gifts or "payments" to birth attendants. Detailed written sources that would allow us to more fully understand what these exchanges meant for the women are not available, but the practice is prominent feature of the Pango women's oral histories that I discuss later in this chapter. In chapter 1, I discussed the ways in which the bride price was a form of exchange between kin groups and a part of social reproduction that became the focus of colonial intervention during a period of depopulation. Colonial concern with population size linked reproduction and economy in ways that overlapped with Ni-Vanuatu concerns of social reproduction and exchange. Here, reproduction and exchange again bring together the concerns of colonial agents and Ni-Vanuatu, however far apart their respective understandings of these practices are. When developing a payment scheme for staff working at the rural clinics that had been newly placed under their control, the NHBS made use of the informal Indigenous system of "paying" the women who assist with childbirth to avoid being involved in paying the midwives in rural areas. The payment scheme for hospital staff shows the colonial concern with economics and reproduction, which made the midwives' "payment" visible to colonial officials, medical practitioners, and anthropologists. It also shows an aspect of how the colonial administration depended on Indigenous social organization in the way that they delivered their projects. Elsewhere in the Global South during this period, biopolitical interventions linking population, reproduction, and economy took the form of invasive population control measures. These measures explicitly rendered life in economistic terms and determined which lives were worth investing in. The context in Vanuatu shows a different dimension of colonial biopolitics during this period. Here, the colonial connection between economy and reproduction was not limited to amplifying the economization of life in terms of GDP. Rendering the non-monetized Indigenous exchanges of reproduction visible for scientific interest and using the networks for colonial systems was another way of extending colonial influence and encouraging the medicalization of birth. The connection between birth, reproduction, and economy (both formal and exchange economies) was where Ni-Vanuatu and colonial officials interests intersected.

Technologies and Material Infrastructures of Birth

The socio-technical systems of biomedicine require adequate physical infrastructure and resources in a "stable place" (Street 2014). Never particularly grandiose, by the 1960s, the facilities at the PMH were in poor repair; in fact, medical superintendents had been fundraising among the faithful in Australia for a new building since at least 1937. As the NHBS prepared to assume management of health services founded by the anglophone missionaries, the plan was to close the hospital on Iririki Island and open one in Port Vila. The foundation stone of the Vila Central Hospital (VCH) was laid on 8 September 1969, and the hospital eventually opened in 1975 (Hawka 2019). The VCH remains the biggest hospital in Vanuatu. Even with the projected opening of a new hospital in the near future, Dr. Freeman, the medical superintendent, demanded improved conditions for birthing women at the PMH in the 1960s. Though the Presbyterian mission had put considerable effort into training nurses and encouraging women to have their babies in the hospital, by the mid-1960s, the facilities at the PMH were desperately lacking. In the archival records, Dr. Freeman's frustration is palpable as he described the woeful conditions for birthing women (and indeed all patients). Dr. Freeman began advocating for a separate maternity ward in 1964 and was still advocating this in 1967.

Dr. Freeman recorded the conditions of "gross overcrowding" and that

> maternity cases are kept in the female ward in contact with pneumonias, ulcerated legs, breast abscesses, kidney infections … all a potent source of infection. On the east verandah, just outside the windows and doors of the General ward, are the children on the floor with ear infections, chest infections, malnutrition, hook worm, malaria and skin diseases. On the west verandah, just outside the windows and doors of the general wards are the dysenteries and the active TB cases, the measles and whatever else of a highly infectious nature has been admitted to hospital … In addition to this, the limited toilet facilities are used by all female patients, and in fact, the puerperal women are obliged to go through the TB and dysentery patients into the same closets which these women use. (Freeman 1964)

He continued to describe the labour ward as follows:

> This is a room 10 feet by 9 feet 9 inches, a total area of 97.5 square feet into which is placed the following:
>
> 1 1 obstetric bed
> 2 1 cabinet containing bowls, gowns, gloves, masks

3 1 sink (the only one supplied with running water in any ward in the hospital)
4 1 instrument table
5 1 table beside the patient for personal articles
6 1 cylinder of oxygen for use in resuscitation of babies
7 1 wall shelf for pads, drapes, bowls and charts

To all of this is added at time of delivery:

1 1 patient and one or more infants
2 1 Doctor or Sister
3 1 Trainee Nurse

As all nurses receive obstetric training in delivery techniques, there are frequent attempts to bring in more nurses for observation. If the cases happen to be abnormal, requiring obstetric manoeuvres, the following must be added:

1 1 other Doctor for administration of anaesthetic
2 Anaesthetic equipment
3 1 electric sucker
4 1 transfusion stand, bottle of blood or intravenous fluid and drip set
5 Extra instruments
6 An extra Sister or Nurse (Freeman 1964)

Dr. Freeman saw this as an utterly impossible situation, unsafe for mothers and infants, and he claimed in his memoir that he refused to continue under these conditions. The material conditions meant that, by the doctor's own declaration, the hospital was unsafe for birthing women, and he advocated on their behalf through various public channels.

Dr. Freeman continually expressed his anger to BRC Alexander Wilkie. And BRC Wilkie, in turn, when forwarding Dr. Freeman's 1964 requests to Dr. W.J.M. Evans, the deputy medical adviser at the Ministry for Overseas Development in London, reported that these conditions had horrified a visiting British member of Parliament who had been sent by the British secretary of state. Dr. Wilkie asked Dr. Evans to support an interim solution when the matter was passed on to authorities in London until the VCH could be built. Recognizing that revamping a maternity ward in a hospital that was soon to be replaced might be a case of throwing good money after bad, Wilkie ended his letter by stating that "I personally feel that the situation is pretty embarrassing and that we should take early steps to remedy it" (Wilkie 1965).

The dearth of biomedical resources, even once the transition from village to hospital birth had taken place, had the potential to create birthing conditions deemed unsafe according to the standards of the medical superintendent of the day. Though unacceptable, it is not particularly surprising to learn of inadequate hospital conditions in a resource-poor context. Part of what I am working to show in this chapter is that attempts to medicalize childbirth were part of the larger effort to bring childbirth into a public realm, where concerned medical staff could advocate for safer conditions as matters of public concern.

Having been exposed to Western goods, people, and technologies of violence to an unprecedented degree during World War II, Pacific Islanders entered the post-war era with high expectations for social change. By 1967, the urban and peri-urban population of Port Vila and surrounding villages was 7,738 (McArthur and Yaxley 1968, 31).[10] Rawlings (1999a) writes that the 1960s and 1970s were a time of dramatic growth in Port Vila, which meant that Pango people had access to wage labour jobs and educational opportunities, while also being subject to greater colonial surveillance. Pango men worked as teachers, groundskeepers, and police officers, in addition to filling a variety of other colonial posts. They worked as clergy, as they had before this period as well. Women had some paid employment; they worked as domestic workers (*haos gel*) for settlers in Port Vila (Rodman et al. 2007), or retail workers, such as at the BP (Burns Philip) store. According to local oral histories, women from Pango started selling food in the Port Vila market. Gardens remained important sources of food and housing materials. By the early 1970s, the New Hebrides had been established as a tax haven (Rawlings 2004), and many people in Pango were employed in the banking industry, even if they failed to earn a solid income there.[11] At the encouragement of the British agent and Pango chief Siel Riaman (Rawlings 1999a, 83), Pango witnessed a growth in the birth rate beginning in the 1950s; this was explained to me as being related to strong penalties for abortion and the general encouragement of large families.

The juxtaposition of the growth of the monetary economy in Port Vila during the 1960s and the decline of conditions at the PMH is noteworthy. Relatedly, the growth of the wage economy in urban areas was documented in the first census (which was also the first time livelihood was enumerated), as I discuss in chapter 3. Despite the growth of the monetary economy, the necessary resources were not being funnelled into public health infrastructure. That the colonial administration would not provide adequate technology and medicine at the mission's public hospital shows which lives were deemed worthy of support; it is also clear that resources were not directed towards these goals from the newly expanding cash economy. What arrived with the colonial attempt to build biomedical institutions for birth was an ongoing need for resources, generally on a scale that required state capital in order to maintain physical infrastructure and provide for professional training. Essential resources for these would always be in short supply. It would take until 2020 for a solid maternity ward to be completed with contributions from Australia and an investment bank, as I discuss in the epilogue.

Thus far, I have focused on written sources to show how bringing birth into the hospital meant working to standardize knowledge and staff payment and advocating for adequate technology and infrastructure. Throughout, the nurses, mainly white nurses, are emphasized as moral figures training Ni-Vanuatu women. I turn now to Ni-Vanuatu women's stories of this time, including the memories of nurses like Madeleine who synthesized biomedical knowledge, learned to use many kinds of technology, and offered care to birthing women in a hospital setting. These nurses were among the first Ni-Vanuatu women trained in a Western educational system and they acted as intermediaries between Indigenous and biomedical knowledge paradigms and forms of social organization associated with birth, fertility, and health. Throughout Pango women's narrations of village births and nursing training, as well as their experience of giving birth at the PMH, new forms of knowledge and birthing practices come together in the nurses' care. Bringing these relationships into analytical focus reveals that Indigenous women were active producers of what might be called an Indigenous modernity, and not merely the object of these interventions. This is what it meant to be a moral figure.

Of Village Birth

Rose was a beloved grandma in Pango. When I arrived at her place accompanied by one of her granddaughters, she invited us to sit in her kitchen, a small shelter with thatch roof and corrugated iron walls with space for a fire. She listened to a description of my research project and said generously, "Your work is BIG!!!" I had come to talk with her because Rose was known as an expert in leaf medicine and massage and also because I had been told that she had been a nurse at the PMH. While the relational care knowledge of female massage healers forms the heart of chapter 5, here I focus on their social place as mediators and conveyors of relational infrastructures of care. We began by talking about Rose's work (in 2010) caring for pregnant women. She said that at about six or seven months of pregnancy, women would come and see her to get massaged so that the fetus's head would turn downward. At nine months, approximately a week before going to the hospital, women came to her to get leaves to help them have a quick labour and to dull the pain. She told me that there are some leaves you use to prepare a tincture with hot water, and some with cold water. The hot one is for a sore tongue, or a cold or a cough. The cold one is for mothers about to have a baby. She insisted that one must collect leaves in the sun, as those picked in shadows are *tabu*. When I asked whether there were leaves for men and not women, she replied that women have leaves for women's problems. She had little to say about men's leaves, except that men have leaves for keeping pigs healthy and to make them grow fat. The leaves have to be prepared by her, or her daughter, otherwise they would not work. The leaves like her hand only, she said, because *Save i folem laen blo mama* (Knowledge is passed down the mother's line).

When I asked how she learned about the leaves, she told me that her grandma showed her when she was as young as nine, and that she knew a lot by about eleven years of age. They would go to the bush and gather leaves together. Rose's grandma also assisted women during birth in the village; this was before women began going to the PMH. Rose described what she remembers about such births. To give birth, a woman would lie down. The midwife would put calico down and make the fire. The midwives would take all the *doti* (dirty stuff or garbage) and take it down to the sea. The woman could not cook or touch other people's food for five days. The family made food. There were leaves to make the uterus (*bel*) go down after the baby was born. Such leaves are no longer needed, she said, because there is the hospital.

Like Rose, Madeleine became interested in how the village midwives cared for women, and so she began following them in the village. Madeleine recounted how women were assisted during their pregnancies by more senior women who were typically taught by their own mothers or maternal aunts because they showed interest when they were young. These women were part of matrilineages that carried the knowledge of plants and medicines. The women would do some prenatal care, generally from approximately the sixth month onward. These experts would massage the sore bodies of pregnant women and ensure that the fetus was positioned with its head down. They were often successful at turning an improperly positioned fetus and they had considerable knowledge of plants for tinctures that women would drink in order to ensure a quick labour.

When Rose says *Save i folem laen blo mama*, anthropologists say that such knowledge has been kept in lineages or clans (*naflak*), passed from mothers to their daughters. *Naflak* are social groupings held together by matrilineal links. They are of decreasing social importance, say many senior people in Pango, and are giving way to kin systems called *blad laen* (bloodline), which favour father-son links. Like Rose, a woman would typically learn this knowledge through an apprenticeship with an experienced older matrilineal kinswoman and then pass her expertise to her female descendants, thereby keeping the knowledge in their *naflak*. The transmission of knowledge is not uniform but shaped by which daughters show interest. Some exceptions were made if a young girl outside the matrilineal line showed particular interest in learning these skills.

Madeleine, having worked much of her life as a biomedically trained midwife, remembered village births more clearly than Rose. According to Madeleine, women would give birth on pandanus mats in a thatch house. The attendants would build a fire to warm the back of the new mother and to help stop the blood after the baby was born. Afterwards, they helped the mother heal and be strong. When these women worked, plastic basins were not widely available, Madeleine said, so the carving of wooden tubs was part of the preparation for birth. They would attend the last part of the labour and the birth and care for the mother during the first post-partum days. In their tool kits, the women would have thread, calico, and a knife from a bamboo stock, which was incredibly sharp. After birth, the two village midwives who had assisted the woman would burn the soiled calico and take the placenta down to the shore, clean it in the ocean, and then bury it at the beach. Once it was established that the mother and baby were healthy, after five days or so, the expert would leave and the new mother would thank her by giving mats (sometimes even painted mats with feathers sewn around the edges), bundles of calico, or other important gifts suitable for ceremonial exchanges that she had prepared before the birth. After this, the mother's female relatives would care for her and the newborn infant, while the mother ideally remained at home for the first month. Madeleine, along with other senior women in Pango, thus described very similar conditions and exchange practices as those documented by anthropologists throughout the archipelago in the first half of the twentieth century.

In Pango, objects like calico, pandanus mats, island dresses, bags of rice or sugar, and pigs are ceremonially exchanged at life cycle events like weddings, funerals, and births, to mend relationships between people and between people and ancestral spirits, or to express gratitude for services that are typically not paid for with cash. When I asked about the meaning of this particular set of gifts, I heard that it was for "saying thank you" or *klinim fes* – that is, to make the relationship between the mother and midwife become strong, as one woman explained to me. I did not ever hear it described as a payment. Madeleine in particular said it was important because the midwife had seen the woman's *tabu* parts during birth. Such perceptions of the female body were part of larger ancestral cosmologies that required seclusion during birth and the post-partum period and gender segregation during menstruation, as these were times when feminine potencies were understood as damaging to masculinities.

Senior women in Pango claimed that village births were common until the late 1940 and early 1950s. The arrival of trucks, which could be used to drive people to the wharf, thereby saving them an arduous walk over a sizable hill, was often mentioned as the reason for this shift. The technical skills of the village midwives are not widely remembered in Pango anymore, except by aging experts like Madeleine and Rose. To the extent that these women's work is commonly remembered by other women in Pango, special focus is given to the fires they would build to care for the women during labour, the prenatal massages, and to a lesser extent, the gifts that would be exchanged afterwards. What came out in their oral histories, though, in the tone and warmth of their voices, was the presence of care for the women.

As Rose's and Madeleine's accounts both show, after giving birth, the new mother would traditionally stay in her house for a month while close female kin cared for her and her baby. The new mother would ideally avoid certain foods and avoid combing her hair. This is seldom observed today. The older women I interviewed frequently expressed regret that this was no longer practised. Madeleine's depictions were typical:

MADELEINE: Before, there were lots of things you could not do. You had to stay home for a month. You could not comb your hair, you had to cover yourself up to stay warm. No shellfish, octopus, or coconut. You had to eat warm food to make the baby grow strong through the milk. You had to make a fire inside to stay warm.

A.W.: But they do that now.

MADELEINE: No, they do not do it like that. They walk around. They go out. [*Spoken in a tone of regret.*]

A.W.: I guess now women cannot do it because they aren't around enough people to care for them. Other women work in town.

MADELEINE: That's right [*Hemia nao*].

These post-partum rituals of the past are a reminder of a time of when women could give and experience care with fewer pressures from wage labour. The past is remembered as a less selfish time.

Remembering Nurse Training

The first medically trained nurse-midwives are no longer alive, but I interviewed several women from the following generation, mainly in Pango, about their training at nursing school in the 1950s. When I first spoke with Madeleine, I told her that I was interested in her memories of her training and nursing work because I had been able to read the doctors' reports, including those of Ni-Vanuatu doctors, but I never knew what the nurses thought. To this she answered wryly, "Yes, the doctors wrote lots of reports." The women would typically start learning the relevant skills as teenagers, and they would stop when they married and had children of their own. Sometimes they did continue working at the hospital rather soon after having children or returned after a hiatus. The training program was a forum in which they learned to use new technologies and comfort measures for assisting birthing women and their infants. Learning to make a bed was prominent in all of their memories. Mary, whose memory is no longer what it once was, above all remembers making the beds with clean white sheets. She gestured slowly with her right hand to show me how she would carefully smooth the clean sheets and tuck the corners tightly under the mattress with a flick of the wrist. Another aspect of the curriculum the women typically recalled was learning to bathe the patients, an important skill as new mothers would typically stay in the hospital for a week after giving birth.

Madeleine recalls that in the first year of her training, the women learned how to make beds, how to bathe sick people in bed, how to take temperatures with a glass thermometer, and how to give needles. In the second year, they were able to give medication according to the doctors' orders and they began to observe births. By the third year, the nurse trainees were helping to deliver babies. They learned techniques for preparing the instruments, basins, and sheets during the final stage of labour, for cutting and tying the umbilical cord, and for safely removing the placenta. They were also taught to offer what nurses today commonly call "comfort measures": to rub the mother's back and encourage her to walk during labour (pain medications were not available for normal births).[12] After the birth, the nurses followed a regime of hygiene for cleaning the new mother, wrapping the baby in calico (cotton cloth), and then giving the baby to the mother to be nursed. The newborns slept in a nursery, and the Ni-Vanuatu nurses brought the babies to their mothers to be nursed at four-hour intervals, as per the nursing sister's orders.

Rose, who did her nurse training at the PMH,[13] has warm memories of the maternity ward. She left nursing because she found the work, especially the long hours, too hard with small children. After talking about her knowledge of leaves, we came to talk about her experience as a nurse at the PMH from 1957 to 1959. As I mentioned earlier, her interest in leaves began around the age of nine, when she would go with her grandma, a village midwife, to the bush and gather leaves with her. This developed into an interest in attending births and in the rudiments of care more generally. We engaged in a meandering conversation about her memories of nursing at the PMH; the following is edited slightly for clarity:

A.W.: Were the women from Pango glad to go to hospital, or did they have to be convinced?

ROSE: Oh no, they were glad to go. If the women couldn't make it up the hill from the launch, then the nurse came down with the stretcher. They gave birth on a bed in labour room. There was a European room and then the maternity ward, where everyone else was. Nurses could stitch up small tears, otherwise the doctor did it. They had anaesthetic.

A.W.: So, what did the women do in labour?

ROSE: Walked around, then, when it got serious, they would go to labour room and lie down. The women would give birth with the nurse.

Rose was proud to say that, back then, the women could stay at the hospital for seven days, not like the twenty-four-hour discharges practised at the government hospital at the time of our interview:

ROSE: The Black nurses [*Blak nas*] would take care of the mamas. They would bathe them. They would change their bed pan. Now, you have to go to the bathroom yourself.

The babies slept in the nursery. The *Blak nas* brought the babies to the mothers at feeding time. Every four hours. There were five or six babies usually in the nursery. There were visiting hours from 8:30 to 9:00. The papas would visit then go back to their place. Nurses slept at the hospital or in the nursing quarters.

A.W.: Were the women scared to go to the hospital?

ROSE: They felt ashamed to let a man look, and especially a white man. So, the doctor would talk to the *Blak nas*, and then the *Blak nas* would explain everything to them so that they wouldn't be scared.

A.W.: Did you have advice for mothers?
ROSE: Eat *aelan kakae* and give it to your baby. Bring your baby to be weighed.
A.W.: Were there things in your nurse training that conflicted with your knowledge about leaves and *kastom*?
ROSE: No. When you have knowledge, it's yours. [*Sapos yu gat save, yu gat save*].

Nurses were part of the introduction of biomedicine elsewhere in the Pacific Islands as well. Beginning in 1907, Indigenous *pattera* on Guam were trained as nurse-midwives by the US Navy; in this way, they served as a link between Chamorro traditional practices of birth and infant care and US hospital-based care. Thus, as Anne Perez Hattori (2006) argues, the Indigenous caregivers made the US Navy hospital into a hybrid space of traditional and modern beliefs and practice. DeLisle (2015) shows beautifully how *pattera* persistance in maintaining Indigenous birthing knowledge and practice was a form of gendered Indigenous resistance to US military occupation. When training Indigenous women to be obstetric nurses on Fiji, the formally trained nurses initially maintained harsh judgments of local midwives who lacked a professional education, but in time a more cooperative relationship developed (Lukere 2002). In postcolonial Papua New Guinea, as Barbara Andersen has shown, nurses mobilize idioms of time in order to assert their authority and "to sort and discipline moral subjects" (2016, 24). Ni-Vanuatu nurses during this time served a similar role as moral figures who embodied old and new forms of knowledge and, perhaps most importantly, care. They stand as emblems of a different moment when there was more time for care, in contrast to the selfishness of the present, dominated as it is by wage labour.

Grandmothers Remembering Hospital Birth

So far in this chapter, I have presented some of the biopolitical measures the British and English-speaking missionaries took to bring birth to the hospital. I then presented former Ni-Vanuatu nurses' oral histories of their training in and experiences of village births, demonstrating the importance of Indigenous social practices as relational infrastructures in how nurses were selected and successful in their work. In this final section, I show how oral histories of women who gave birth at the hospital in the 1960s focus not on the technology or knowledge of the nurses, but rather the care they provided. The histories also show how the organization of work and care has changed in the last four decades.

Francine was a powerful woman in her early seventies. She was taking care of her husband, who was forced to use a wheelchair due to complications from diabetes, her unmarried brother-in-law, and several of her children and grandchildren, who ran in and out of her house. When we first met, she sized me up and said with a laugh, "If I tell you what I know, your book will be VERY big!" I swallowed and, for neither the first nor the last time, fought back doubts about whether I was up to the task. Over a series of interviews at the table at the front of her sizable cinder-block house at an edge of the village, she taught me a great deal about the history of Pango, and local women's lives in particular.

While the colonial administration's recruitment of men for medical school in Suva was based on candidates' ability to read and write (Widmer 2010), the missionary running the hospital took a different approach to enrolling nurse-midwives prior to the beginning of formal training. Francine recalled that the PMH doctor contacted Reverend Mackenzie, the Presbyterian missionary living in Pango, and invited him to suggest women who could train as nurse-midwife apprentices at the hospital. Mackenzie, who knew the village well, named three women who were already attending births in the village. According to various oral histories, these were women who already had ancestral knowledge, and several, though not all, of the first nursing trainees had prior knowledge and status in the village. Furthermore, Francine told me with authority that women from three villages around Port Vila – Mele, Erakor, Pango – would take turns bringing food to the hospital. In summing up her memories of the hospital, she said, "The PMH was good. The nurses were good. They were just like us."

Other women who had given birth at the PMH had similar memories of the hospital. For example, Olive said with a smile,

> At the PMH, the nurses got us to do exercises. The mamas would sit on their beds and move their arms. They would be worried about going to the bathroom or peeing themselves, but it was to make the body get strong. We prayed every morning and every morning they did exercises. All the mamas in two rows. The Black nurses [*blak nas*] made us do it too. The exercises made me feel good.

The hospital of the past is remembered as place of care, and such memories were frequently contrasted with the reality of 2010, by which point families were forced to bring their own supplies to the hospital. Ni-Vanuatu nurses were key in women's narrations of the hospital environment as a place of care.

Women's stories of birth at the PMH were enmeshed in their own life histories. These life stories followed a common arc. The women would begin their stories with where they lived as children. Then they moved to telling about the time when they were figuring out who their husband would be, which meant a discussion of whether their marriage was arranged or not, and then where they would live after marriage. Given that the ideal practice in Pango was for a woman to go live with her husband on his family's land after getting married, if the husband's family was from elsewhere on Efate or Vanuatu, and the woman wanted to stay in Pango, the women had to negotiate land from her own family, usually her father. The women would continue their life histories by recounting pregnancies and births, the experience of raising children, and then seeing their own children get married. The stories tended to end with the women smiling and saying, "Now I can have a little rest" (*spel smol*).

Oral Histories and Histories of the Present

While histories can often tell us as much about the present as they do about the past, this is particularly true for oral histories and oral life histories. Such histories, of course, reflect a particular kind of situated knowledge, and need to be interpreted with attention to their local meanings and in concert with available archival sources. First, the women's tendency to emphasize the good qualities of the missionaries and downplay any of the shortcomings of the PMH in their recollections is in keeping with the sort of Christian comportment thought to befit an older woman in Pango. It would be impolite to directly complain about overseas missionaries and their institution in an interview with a researcher from Canada. Furthermore, especially for the oldest women I interviewed, expatriate missionaries in general (extending back to the 1800s) are remembered with affection. I listened to these histories around the thirtieth anniversary of Vanuatu's independence, a time of public reflection on what had been accomplished since 1980. There was a great deal of what might be called postcolonial discontent at what the government had managed to deliver in the way of healthcare services. For example, people in Pango wanted a clinic in their village, like the one run by Madeleine before independence, and were disappointed that they had to go into town for those services, which themselves were substandard.

The way that these women look back on birth at the PMH contrasts with how they describe birth at the current hospital, not only because of the differences in technology or infrastructure, but especially because of the differences in care at the hospital and care networks in the village now. Madeleine's descriptions were particularly articulate versions of what I frequently heard. In one of our conversations, I asked her about the difference between giving birth at the mission and the government hospital. Madeleine explained that "at the PMH, everything was provided. You didn't need to bring anything. The hospital had calico, blankets, food. Now, at the government hospital, there is a list of things that you have to bring when you have your baby. You have to bring everything. You need to buy it yourself." The past was characterized by care and community; the present by selfishness and the need to earn money in town.

I see the way that these senior women recount the past as a critique, not of the fact that birth has become the purview of the hospital, but that the networks of care and social reproduction have been diminished. Village midwives were remembered for the care they provided, in the form of lighting fires, massaging painful bodies, administering leaf tinctures, bringing the afterbirth and soiled mats to the ocean, and participating in ceremonial exchanges. In recounting the nurses at the PMH, women's memories of the care they received were associated with the Christian values the nurses embodied and the biomedical technologies they provided. These were evidenced in their changing of bed pans, leading of exercises, bringing babies for feeding, changing beds, leading prayers, etc. Post-partum care was recognized in the women who cared for the new mother in the village while she cared for the infant for a month.

In these narratives, care and technologies are juxtaposed with the present, in which senior women and retired nurses point to gaps separating the care they received both at the PMH and in the village and what young women receive (and provide) today. Public care in the postcolonial government hospital and post-partum care in the community is not filling the need for care that has resulted from larger numbers of young women taking up wage labour in town. The past, and particularly the care once provided in both village and hospital, is remembered fondly in light of a present in which the prior traditions and networks of care have been stretched thin.

The time that I was asking the women to remember, the 1950s and 1960s, saw the expansion of the monetary economy and wage labour in Port Vila. Three villages around the capital became peri-urban places from which people could easily commute to town while maintaining some subsistence crops and engaging in various forms of village sociality. Philibert (1988) wrote of this process during a slightly later period (1972–83), when women's participation in wage labour expanded in the nearby village of Erakor. This meant a decline in arranged marriages and a decrease in the importance of gardening because of the lack of time people had to devote to this activity. For the expansion of women's wage labour to happen, and for people to be away from their responsibilities in the village so as to perform wage-paying work, certain social arrangements had to be in place. It required, more specifically, networks of support shaped by kin groups or the *naflak* that were reproduced through daily intergenerational contact. Pango women remember this time as a period when the village was smaller and houses were much closer together and arranged in a circle around the village well. One woman, in what was perhaps an act of bravery, recounted that she did not like the hospital. "The food was bad," she said, and "I wanted to get away from that place, it was noisy." But she nonetheless described, in our conversations about that period of history, her regret about the selfishness and absence of care in the present, and about young women's lack of interest in women's traditional knowledge.

Women's memories of giving birth and the care they received from the nurses at the hospital were thus interwoven into how they told their own life stories. Their life accomplishments – and especially their finding a place to live in Pango and raising healthy children who are now themselves married – in fact represents a lifetime of successful relationships, care, and work. They are successful lives in which the women had met their intergenerational obligations. In this sense, their memories of giving birth and the care they received from the nurses at the PMH form part of a life narrative and village history in which the past is perhaps idealized as a time when obligations within kin groups were met. The 1960s represented a time when the *naflak* and associated care economies were strong (because knowledge was passed through a *naflak*).

At the time of the interviews, each of the women I spoke with lived in robust networks of care in which they, too, had ongoing responsibilities. Their grandchildren either came in and out of their homes regularly or lived there on a permanent basis. Often, their house was the most substantial of their kin group and a place where their families would gather. In general, these Pango women have access to more resources, cash and otherwise, than most women in Vanuatu (and many men in Vanuatu as well), compared to whom they had benefited from the transition to a cash economy. Still, although they might not leave their homes and surrounding compounds very often (meaning that family members would bring them food and other necessities from the garden and the stores in town), the care they provided to family members on a daily basis was important. And care is an important part of the way they narrate the difference between the past and present.

Conclusion

Nurses are those people, Alice Street (2016, 334) writes, who are valued because they care for strangers as though they were their kin. The Ni-Vanuatu women, like Madeleine, who can still give a relaxing prenatal massage or give leaf tinctures learned from female kin, who might remind pregnant women to take their iron and anti-malarial pills from the clinic, and would definitely say a prayer for the well-being of a sick child, are ideal examples of how women in Vanuatu were able to combine biomedicine with Christian values and ancestral techniques to care for kin and others in the clinic and village. They were trained by kin and clinicians and are remembered as moral figures for the care they provided, just as the women who trained them in village birth are remembered for the care they gave, the fires they tended, and the backs they massaged during painful labours.

I have presented aspects of Pango women's oral histories here as they are interwoven with the history of the shift in childbirth from village to mission hospital that was part of the gradual centralization of control over hospitals under colonial administration. This might appear like a transition whereby Indigenous practices were replaced by modern medical care offered in Christian service, and finally secular state medical health care, but this is not a linear narrative of progress. The Indigenous and biomedical practices, as well as the Christian values, that Ni-Vanuatu nurses draw upon in their care for pregnant women coexist; indeed, this ongoing negotiation of Christian, Indigenous, and external knowledge (like biomedicine) is at the very core of what modernity means in Vanuatu. Ni-Vanuatu nurses navigated this moral terrain and stand as moral figures in people's memories of the PMH.

At the time of our conversations, senior Pango women were living with the reality of the shift from Christian to state hospital and to the increasing participation of women in wage labour that began in earnest in the 1960s. Their fond memories of the moral figures of Ni-Vanuatu nurses, which were generally juxtaposed with later experiences of neglect at the government hospital and the reduction of traditions that require female care in the village, can be read as postcolonial discontent at the deferral of the long-promised dividends of modernization, and particularly the perception that the state has failed to provide collective care.

This chapter has shown how the expansion of hospital birth in Vanuatu was accomplished in biopolitical terms through the creation of a moral figure alongside the expansion and standardization of biomedical knowledge, techniques, and infrastructure, a process motivated by a desire to maintain a healthy population. But it also shows how this was accomplished through the social networks and care economies of Ni-Vanuatu in which these moral figures, Ni-Vanuatu nurses, were located. The experience of giving birth at the hospital is remembered fondly, even though the technology was limited, because of the care of the nurses. Memories of the mission hospital are contrasted with the experience of diminished social networks of care in the postcolonial context, both at the hospital and in daily life, where they are reduced and displaced as a consequence of the increasing dominance of wage labour, particularly among young women.

The experience of feeling invisible at a postcolonial hospital presents a powerful contrast with the global expansion of the biopolitics by which reproduction was medicalized and in which a key objective was rendering people and life processes visible in a population (statistically, demographically, epidemiologically). In Vanuatu, colonial officials and missionaries demonstrated concern for improving the mortality rate and overall health of the population by training and paying educated workers and finding capital to build and maintain infrastructure. Acquiring, using, and advocating for biomedical technologies and infrastructures were partly how reproduction became a public concern. Furthermore, in post-war Vanuatu, the infrastructures of medicalization were entangled with economic thinking, but not with an overwhelming concern for GDP. Rather, medicalizing reproduction meant making unpaid care labour and non-monetary exchange visible and depending on them as relational infrastructures in order for the technological aspects of the infrastructure to work.

During the 1960s, the colonial project meant expanding training and standardizing knowledge for nurses. Looking at how birth caregivers were made visible to and by the colonial project shows how reproduction and economies, both monetary and otherwise, were connected in the biopolitical aspects of colonial governance, as we saw in the discussion of archived documents in the first half of the chapter. Reproduction in this colonial context, as a privileged object of both biomedical practice and colonial governance, depended on Indigenous networks and forms of sociality, and infrastructures of care economies in particular. The labour, knowledge, and medical technologies through which colonial biopolitics are generally seen to proliferate show how reproduction is a distributed process that depends on Indigenous care economies and social networks.

Chapter Three

"It Will Help Planning for the Future": Making Men's and Women's "Subsistence" Public Knowledge in the First Census, 1966–1967

Land to Ni-Vanuatu is what a mother is to a baby. It is with land that he defines his identity and it is with land that he maintains his spiritual strength. Ni-Vanuatu do allow others the use of their land, but they always retain the right of ownership.

Hon. Sethy Regenvanu, First Minister of Lands (quoted in Van Trease 1987, xi)

In any village with a radio, Ni-Vanuatu would have heard the census song broadcast over the airwaves in the months leading up to 28 May 1967:

Long May No. 28, Census Day 'e come
Some teacher 'e go walkabout, 'e lookout every man
He askem every somet'ing, b'long all 'e tell out good
Long May No. 28, Census Day 'e come.

Long islands long New Hebrides, how much man 'e stop?
M'bye teacher 'e askem every man, how much wife 'e got?
How much pikinini, 'e askem everyone
Long May No. 28, Census Day 'e come.

Long May No. 28, some teacher 'e lookout you
'E askem long every man, how much year 'e got
'E askem how much year long every pikinini too
Long May No. 28, Census Day 'e come.

Long May No. 28, Census Day 'e come
Some teacher 'e come look you, 'e makem plenty talk
You tell long place where you you born, more place where you you work
Long May No. 28, Census Day 'e come.[1]

A string band recorded this census song at the behest of Australian demographer Norma McArthur and her team, and it was broadcast widely on the radio. The song followed the tune of an Australian folk song used in a 1966 Australian census jingle (McArthur and Yaxley 1968, 74). In Vanuatu, the grand undertaking to quantify life across more than eighty islands and to make social and biological reproduction visible for public policy took place after roughly a century of researchers, medical doctors, and missionaries complaining to the colonial authorities about the lack of reliable demographic data with which to measure population trends. In 1961, the BRC was still fielding enquiries about population statistics with awkward ignorance. Institutions like the *Encyclopaedia Britannica* asked after the "general population census" and "censuses of agriculture, livestock, industry and commerce" (Martin 1961), while in 1966, the British Ministry of Overseas Development asked for census data for the Colonial Office's "Digest of Statistics" (Ministrant 1966).

By the time the census song was being broadcast across the airwaves, colonial authorities had hired demographer Norma McArthur[2] (1921–84), together with the French administrative officer Mr. J. Fabre and British administrator John Yaxley, to supervise the first simultaneous census of the New Hebrides. How to motivate people to participate was an important concern for the administrators, and the census song was prepared and broadcast with this aim in mind. The teachers mentioned in the song were Ni-Vanuatu and had been trained for up to three days on how to work as enumerators. As I will discuss, they were trained to say that the purpose of the census was "to help the government make plans," a benefit the census administrators presented as self-evident and which they assumed would compel people to answer their questions.

The census is a key technology that governments use to categorize populations and thereby develop social policy and governance strategies. Statistics derived from census data are the bedrock of the forms of population-level biopolitics that Foucault has resoundingly shown to be one of the two poles for the operation of modern biopower. Censuses are significant for how they match social identities with population categories, effectively "making up people" (Hacking 2006) and cementing otherwise fluid terms because one individual cannot be in more than one identity category (Anderson 2001).

The number, so closely associated with the identities of the census, has long been shown to be central to the colonial imagination (Appadurai 1993; Cohn, 1987) and the production of colonial governmentality. The census is part of changing social conditions in ways that "do not reflect a simple expansion of the range of individual choice, but the creation of conditions in which only new (i.e. modern) choices can be made" (Asad 1992, 337). Examining colonial governmentality means being attentive to the following questions, framed succinctly by David Scott (1995, 197):

> In any historical instance, what does colonial power seek to organize and reorganize? In other words, what does colonial power take as the target upon which to work? Moreover, for what project does it require that target-object? And how does it go about securing it in order to realize its ends? In short, what in each instance is colonial power's structure and project as it inserts itself into – or more properly, as it constitutes – the domain of the colonial?

Engaging with Scott, Nira Wickramasinghe writes that colonial governmentality, in Scott's sense of producing a public sphere subject to a colonial rationale, conveys an impression that is far too neat and tidy and grants too much power to the colonial process. Rather, Wickramasinghe argues that in the context of British Ceylon, "anyone who has read minutes of colonial administrators at the Public Records Office will agree that most of them did not have the faintest idea what they were doing ... The quality of colonial rule is better described as haphazard and tentative" (2015, 101). In pre–World War II colonial Vanuatu, this portrayal could also characterize colonial officials in that knowledge desired for governance was generally elusive (Widmer 2008), and these officials often appeared to have an inadequate understanding of Ni-Vanuatu societies. Colonial social policies at that time, such as they were, explicitly deployed persuasion to promote participation in colonial projects. The 1967 census marks a shift in this respect. It was a successful endeavour that provided a public account of important aspects of population dynamics, like birth rate, occupation, and migration.

The census has been central to the categorization of "race" and "ethnicity" and, relatedly, to the enumeration of "migration" in association with state projects. This has especially been the case with colonial censuses and part of the shifting, and yet ongoing, connection between population thinking and race thinking. The population in Vanuatu was categorized this way, with the addition of "Island" as both an identity and geographical location. However, given that most inhabitants were considered "Melanesian," other population-related categories, like "livelihood" and "fertility" rate, took on a larger significance in the 1967 census.

Documenting occupational categories is another of the bureaucratic tasks that censuses perform. An administrative problem faced by the colonial authorities in Vanuatu was the fact that much labour was not monetized. The collection of activities here was not quite the same as "unpaid labour" or "reproductive labour," considered so crucial to capitalism, as Marxist and Marxist feminist scholars have documented, because the unpaid labour in Vanuatu required land. In 1902, a French census had recorded the number of hectares under cultivation by British or French settlers as a way of showing that the scale of French influence was greater than that of the British settlers (Widmer 2017), but how Indigenous people lived in relation to land and labour had not yet been quantified in a census. "Subsistence" was the category in which the census administrators gathered such information.

In this chapter, I show how census categories, and especially the figure of "subsistence,"[3] transformed the biological and social aspects of reproduction into public knowledge. The chapter shows how quantifying reproduction and social reproduction were part of how colonial governmentality worked at this time. This census tracked basic household sizes, the number of births per woman, and, significantly, paid and unpaid livelihoods. What is of note, as I will show, is the fact that the categories needed to be taught in order to be recognized, indicating that they were not self-evident. I illuminate the challenges faced by the planners in their attempts to accommodate Indigenous social forms into their categorization of "subsistence" within census questionnaires, connected as it was to Ni-Vanuatu exchange practices of social reproduction.

"Subsistence" emerges as a historically specific category of knowledge that was aimed at quantifying life and making the non-monetized reproductive labour of women and men publicly visible in particular ways. Through a focus on "subsistence," women were rendered almost as economically active as men. Practices that quantified life through knowledge production and the census demonstrated planners' interests during this time. Generating the category of "subsistence" as something to be planned for and governed was part of how the biological and social dimensions of reproduction were made visible in concert with the economic agendas of "modernization and development" that prevailed across the Pacific, and indeed throughout the world, during this time.

The aspects of reproduction that constitute the figure of "subsistence" cannot be conceived of without access to land. As we saw in the comment from Sethy Regenvanu at the beginning of the chapter, land occupies an absolutely central position in Ni-Vanuatu lives. Indeed, land is never just a means of generating the materials for physical survival – it is part of a moral universe. As I will show, the moral and political dimensions of land and belonging, though absent from the census process, permeate politics in Vanuatu. In the 1960s, such politics took an urgent tone. While the census worked to render social reproduction visible to planners separately from land politics, my intent is to bring land politics into the same frame to show that reproduction, in all its social and biological dimensions, is a distributed process that extends far beyond human bodies.

In this chapter, I show that the public quantification of life within the census rendered men's and women's work and care legible to the state's development agenda during the post-war era, conscripting them as subjects to be planned for in particular ways. I further show how subsistence, conceived as a category of livelihood, is derived from negotiations over what constitutes "work" in relation to the use of cash and waged labour. However, local economies, especially around copra production and exchange, still mattered in terms of the forms this knowledge took. Highlighting the local networks of "subsistence" is yet another way for me to show that reproduction is a distributed process. As we see in the other chapters of this book, when reproduction is made public, it is connected to capitalist and state agendas of the day, development and wage labour in particular. Ni-Vanuatu relationalities and exchange obligations, while they align at times with development and state agendas, also exceed the monetized goals of those plans.

Accordingly, I argue that the quantification of lives in association with their non-monetized labour capacities served to silo birth rates and population concerns separately from land rights. Reproduction, in its biological and social dimensions, was made public separately from changes in land use. Demography and census data, with their focus on counting and categorizing people, led to a focus on births and women's reproduction – a population pressure that would in turn put pressure on the land, rather than a focus on people's reduced access to land because of changes in industry and land use.

Beyond the solidification of identity categories, another effect of censuses was how, during the data-collection processes, as well as in the subsequent collation of the data into tables, relationships were produced between categories that appeared self-evident (like "New Hebridean" and "subsistence"), whereas other possible relationships are silent or eclipsed (like "men" and "birth rates"). While the census song quoted above portrayed the census merely as the task of counting reality, the practices of categorization, calculation, and representation were actually about making complicated processes public in ways that made them appear governable.

Norma McArthur's demographic focus on population structure and fertility rates meant that when reproduction and subsistence were made publicly visible, they were treated separately from land use and land access. Birth rates and population growth were calculated in order to plan for a wage labour force and social institutions that siloed these governmental concerns from land use politics. I attend to these colonial filing and categorization practices with the aim of showing, as was done in chapter 1, that certain practical concerns – say, the ability to hold a single census card that could record as much information as possible – were central to colonial bureaucratic governance. I begin with a consideration of McArthur and her work before turning to an examination of how the census categories were negotiated and taught to Ni-Vanuatu enumerators. I then discuss what is not mentioned in the figure of "subsistence" measured in the census: the land issues that emerged as a political catalyst for Ni-Vanuatu movements for social justice at that time.

McArthur in the Pacific Region

Norma McArthur (1921–84) was trained in biological statistics in the Department of Eugenics, Biometry and Genetics at University College London by the well-known biologists John Haldane and Lionel Penrose. She began an assistant lectureship in demography at the same institution, but when she took a position at the Australian National University in 1952 she embarked upon an intellectual path that would shape her career for decades to come (Obituaries Australia 1984). In the 1950s, she conducted landmark censuses in Fiji, Tonga, Samoa, and the Cook Islands, and later facilitated the design and implementation of censuses in other island territories, namely, the British Solomon Islands in 1959 and the Gilbert and Ellice Islands (now Kiribati and Tuvalu) in 1963. Thus, between 1950 and 1970, many people, most of them Indigenous, living in the Pacific Islands were made statistically legible through official censuses for the first time. McArthur's work became a benchmark for demographic practice in the Pacific Islands and she became known, regionally, as the foremost demographer of her era.

With her growing reputation in the region, the South Pacific Commission (SPC)[4] asked McArthur to write a handbook for colonial administrators and social scientists in the South Pacific; the result was *Introducing Population Statistics* (1961). The SPC was founded in 1947 with a mission to coordinate and fund research to "better the lives of Islanders" under the administration of Australia, France, New Zealand, the Netherlands, the United Kingdom, and the United States (Smith 1972, 28). *Introducing Population Statistics* is a clear and methodical "how-to" volume spanning 137 pages. In it, McArthur (1961, 134) recommends a handful of books for further reading on population statistics, including *Handbook of Population Census Methods* (1958) and *The*

Determinants of Population Trends (1953), both of which were published by the United Nations, as well as *The Measurement of Population Growth* (1935), by Robert Kuczynski, a leading British demographer, who lectured at the London School of Economics and consulted for the British Colonial Office. Through her selection of this short list of publications, she invokes the global governance and health surveillance mechanisms prevailing within contemporary British colonial and international institutions keen on modernization and development.

Intended for colonial governments, the advice McArthur provides in this manual includes practical tips for designing censuses and surveys. For efficient data collection and analysis, she recommends that the forms that enumerators carry on the day of the census be organized so as to hold information from one household per form, and that each form should be sufficiently large to contain information for all household members on one card (the forms were 17 × 13.5 inches in size). She writes that the design and layout are crucial for ensuring accuracy, and provides the following recommendations for the design of census questionnaires:

> A) It should be easy to handle in the field, easy for both the interviewer and the respondents;
> B) the only questions to be asked should be those which are strictly relevant and to which reliable and pertinent answers can be expected; and
> C) these questions should be set out in some fashion convenient to the interviewer and to the subsequent processing. (McArthur 1961, 79)

McArthur (1961, 77) includes a model census card in her manual, a "Household Schedule" that had been used in the Fijian census conducted in 1956. Writing about the design of such cards, she notes that, "the personal particulars which were recorded on these schedules probably represent the maximum which can be obtained with any degree of accuracy in a full enumeration of any Pacific territory at present" (76). As seen in figure 3.1, the largest column and the size of the spaces were carefully designed in order to anticipate the widest possible variety of answers to the following question (number 11): "What work does this person do? For whom and where does he do it?"

She also makes a few suggestions about how to improve the schedule: "Remove the description of the dwelling and redesign the geographic identification, so that two more people can be fitted on the schedule" (76). These suggestions were incorporated in the New Hebrides Household Schedule (see figures 3.2 and 3.3).

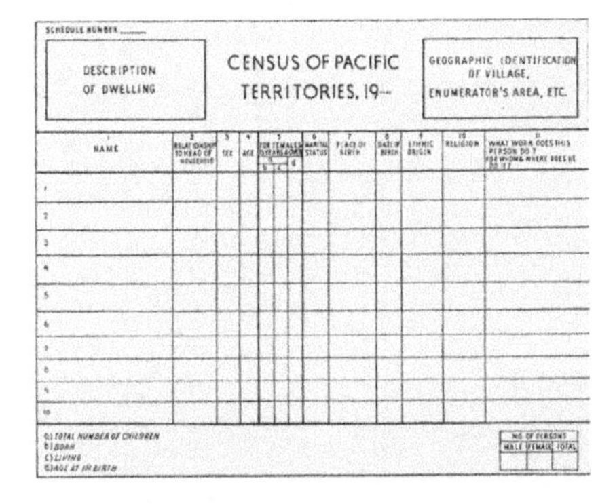
SCHEDULE NUMBER ______

DESCRIPTION OF DWELLING

CENSUS OF PACIFIC TERRITORIES, 19--

GEOGRAPHIC IDENTIFICATION OF VILLAGE, ENUMERATOR'S AREA, ETC.

1 NAME	2 RELATIONSHIP TO HEAD OF HOUSEHOLD	3 SEX	4 AGE	5	6 MARITAL STATUS	7 PLACE OF BIRTH	8 DATE OF BIRTH	9 ETHNIC ORIGIN	10 RELIGION	11 WHAT WORK DOES THIS PERSON DO? FOR WHOM & WHERE DOES HE DO IT?
1										
2										
3										
4										
5										
6										
7										
8										
9										
10										

a) TOTAL NUMBER OF CHILDREN
b) BORN
c) LIVING
d) AGE AT 1st BIRTH

NO. OF PERSONS		
MALE	FEMALE	TOTAL

Figure 3.1. Sample Household Schedule used in Fijian census conducted in 1956
Source: McArthur (1961, 77). Photo courtesy of the author.

Despite her point that more information could not successfully be collected on one card, she added two additional columns to the New Hebrides Household Schedule, "occupation" and "employer and place of work." She elaborated on her rationale in her final report for the New Hebrides census:

> The classification of industry and occupation which has been followed in this census was based on that used in the 1956 Census of Fiji, but this was modified considerably to reflect the less sophisticated economic structure of the Condominium. The word "Industry" is used here to cover all forms of economic activity extending from pure subsistence with no involvement in the cash economy to the relatively specialized professional services. (McArthur and Yaxley 1968, 57)

This addition is significant in that it rendered "subsistence" labour activities visible (a point to which I return later in this chapter). Without it, most Ni-Vanuatu would have been rendered poor, penniless, or lazy; instead, they were rendered visible as active subsistence gardeners whose work and care were recognized as vital dimensions of social life.

In contrast to the practical instructions provided in *Introducing Population Statistics*, in McArthur's (1967a) book titled *Island Populations of the Pacific*, she takes an analytical approach to censuses conducted in Fiji, Tonga, Samoa, the Cook Islands, and French Polynesia to "establish population sizes, the subsequent changes in size and structure and the reasons for these changes" (Campbell 2006, 99). In this benchmark monograph, she focuses on key demographic measurements, namely, fertility rates and the age structure of the population.

NAME OF ISLAND NOM DE L'ILE	1 & 2
ENUMERATORS AREA CODE CODE DE LA ZONE DU RECENSEMENT	3 & 4
NAME OF VILLAGE OR REGION NOM DU VILLAGE OU DE LA REGION	5 & 6
SCHEDULE NUMBER NUMERO DE L'ANNEXE	7, 8 & 9

CONDOMINIUM OF THE NEW HEBRIDES
CONDOMINIUM DES NOUVELLES HEBRIDES
CENSUS 1967 RECENSEMENT

This **HOUSEHOLD SCHEDULE** is to be distributed or completed by an ENUMERATOR for each HOUSEHOLD in the AREA for which he or she is responsible.

Cette **ANNEXE CONCERNANT LES FOYERS** doit être distribuée ou remplie par un RECENSEUR pour chaque FOYER dans LA ZONE dont il ou elle est chargé (e).

CONFIDENTIAL/CONFIDENTIEL

Number of persons in this Household / Nombre de personnes comprises dans ce Foyer		
M	F	TOTAL

1	2	3	4	5			6	7	8	9	10	11	12
NAME NOM	RELATIONSHIP TO HEAD OF HOUSEHOLD e.g. wife, son, visitor, etc. / PARENTÉ AVEC LE CHEF DU FOYER p. ex. épouse, fils, visiteur, etc.	SEX SEXE M or/ou F	AGE LAST BIRTHDAY if under 1 year write "0" / AGE DERNIER ANNIVERSAIRE si moins d'un an, écrivez "0"	FOR FEMALES AGED 15 YEARS OR OVER / POUR LES FEMMES AGÉES DE 15 ANS OU DE PLUS: Total number of children/Nombre total d'enfants: Born/Nés	Still living/Vivants	Age when first child born/âge lors de la naissance du premier enfant	MARITAL STATUS NM = Never Married M = Married W = Widowed D = Divorced / SITUATION MATRIMONIALE JM = Jamais Marié (e) M = Marié (e) V = Veuf (Veuve) D = Divorcé (e)	PLACE OF BIRTH For New Hebrideans Name of island For others name of country / LIEU DE NAISSANCE Pour les Mélanésiens N.H. Nom de l'Ile Pour les autres : nom du pays	LEGAL STATUS NH = New Hebridean B = British or British Ressortissant F = French or French Ressortissant / STATUT LEGAL MNH = Mélanésiens N.H. B = Britannique ou Ressortissant Britannique F = Français ou Ressortissant Français	ETHNIC ORIGIN e.g. New Hebridean, European, Part-European, Vietnamese, Chinese etc. / ORIGINE ETHNIQUE p. ex. Mélanésienne, N.H. Européenne, en partie Européenne, Vietnamienne, Chinoise, etc.	RELIGION Presbyterian, Roman Catholic, Custom, etc. / Presbytérienne, Catholique, Coutumière, etc.	OCCUPATION What work does this person do? / EMPLOI Quel travail fait cette personne ?	EMPLOYER AND PLACE OF WORK Who does he work for and where? / PATRON ET LIEU DE TRAVAIL Pour qui et où travaille-t-elle ?
12		13	14 15	16 17	18 19	20 21	22	23 24	25	26	27 28		29 30 31
1	Head/Chef												
2													
3													
4													
5													
6													
7													
8													
9													

Figure 3.2. Census schedule, New Hebrides, 1967
Source: McArthur and Yaxley (1968). Photo courtesy of the author.

ɹM OF THE NEW HEBRIDES
ᴎ DES NOUVELLES HEBRIDES
1967 RECENSEMENT

CONFIDENTIAL/CONFIDENTIEL

Number of persons in this Household		Nombre de personnes comprises dans ce Foyer
M	F	TOTAL

ↄLD **SCHEDULE** is to be distributed or
IMERATOR for each HOUSEHOLD in the
ɔr she is responsible.

CONCERNANT LES FOYERS doit être dis-
ar un RECENSEUR pour chaque FOYER
l ou elle est chargé (e).

IMPRIMERIE NOUVELLE D.TARDIEU

7		8		9		10		11		12	
PLACE OF BIRTH For New Hebrideans Name of island For others name of country	LIEU DE NAISSANCE Pour les Mélanésiens N.H. Nom de l'Ile Pour les autres : nom du pays	LEGAL STATUS NH = New Hebridean B = British or British Ressortissant F = French or French Ressortissant	STATUT LEGAL MNH = Mélanésiens N.H. B = Britannique ou Ressortissant Britannique F = Français ou Ressortissant Français	ETHNIC ORIGIN e.g. New Hebridean, European, Part-European, Vietnamese, Chinese etc.	ORIGINE ETHNIQUE p. ex. Mélanésienne, N.H. Européenne, en partie Européenne, Vietnamienne, Chinoise, 'etc.	RELIGION Presbyterian, Roman Catholic, Custom, etc.	Presbytérienne, Catholique, Coutumière, , etc.	OCCUPATION What work does this person do?	EMPLOI Quel travail fait cette personne?	EMPLOYER AND PLACE OF WORK Who does he work for and where?	PATRON ET LIEU DE TRAVAIL Pour qui et où travaille-t-elle?
	23 24		25		26		27 28				29 30 31

Figure 3.3. Census schedule: Note column width for questions 11 and 12
Source: McArthur and Yaxley (1968). Photograph courtesy of the author.

These two publications are instructive for understanding how McArthur would approach the New Hebrides census both practically and analytically. In particular, they reveal her methodical and methodological approach, focused as it was on counting people in households organized around male heads, with an emphasis on the kinds of work performed. Her emphasis on births per woman to calculate fertility rates, and ultimately to generate population numbers, resulted in data that would be used to plan social services for particular segments of society. This focus on births per woman also satisfied McArthur's interest in the biological dimensions of population structure after population decline.

McArthur's focus, as a demographer interested in scientific knowledge and practical applications, was on the Pacific Islands' rapid population decline in the late nineteenth century and the first decades of the twentieth century and its enduring impact. In this, she sidestepped some of the more prominent demographic concerns of the global post-war context. Greenhalgh (1996) argues that post-war demography was increasingly part of American Cold War foreign policy, given the prevailing perception of population growth as a threat to peace and a potential contributor to the spread of communism. In many places in the world, such as Nepal and India, fertility reduction was a primary objective of census collection and demographic analysis. This was also a time when demographers tied population growth to environmental degradation in their analyses. The view that global food production was connected to population growth in many parts of the developing world was met with critiques from the Left, which attributed hunger to economic inequality and distribution rather than a lack of food production (Robertson 2016, 224–5). Family planning programs, conducted in the name of modernization and development – themselves key concerns of this time – were carried out by an array of actors, including the foreign aid organizations created by various governments (Dörnemann and Huhle 2016, 142). Notably, the American population experts Kingsley Davis and Frank W. Notestein had revived and reformulated "demographic transition theory" as a central framework of demography during this period. In their widely endorsed scheme, the countries of the globe were divided according to their development statuses and could transition from traditionally high fertility rates to modern (i.e., low) rates with appropriate modernization and biomedical family planning.

In her conclusion to *Island Populations of the Pacific*, McArthur emphasizes that her central interest is the causes of depopulation as they relate to Pacific Islanders' contact with Europeans and the long-term effects of such contact on the population structures of the region. She attributes the population increase she observed in the censuses she conducted in other Pacific Islands to declining mortality, and only small proportions of this growth to increased fertility. Throughout her work, she focused on the practical need to collect reliable data (the absence of which had long been a concern in the Pacific) and not on the more central themes of the discipline foregrounded in European and American settings. These emphases can be seen in the census categories that she taught people to enumerate in the New Hebrides and those matters that became logistically and governmentally separated from population statistics and concerns, namely, land access and land use politics.

Undertaking the First Comprehensive Census

Censuses entail enormous amounts of practical, technical, and scientific labour. In the New Hebrides, there were particular challenges associated with the effort to reach consensus among the French and British authorities over the methods, scope, and cost of the census.[5] In 1962, the British administration had asked Norma McArthur to consider undertaking a census of the archipelago, but this request met with resistance from French administrators. Finally, in 1964, a British official, M. Townsend, wrote to McArthur asking her to again consider the 1962 request. In his letter, Townsend described a debate on economic development that had taken place at a recent meeting of the New Hebrides Advisory Council. The council (formed in 1957) was the main mechanism enabling Ni-Vanuatu to participate in state politics. Ni-Vanuatu were legally stateless during the years of the Condominium, having no rights to British or French citizenship. By 1964, Ni-Vanuatu held ten of the council's twenty-six seats (the corresponding ratio at the council's founding was four of sixteen seats; Jackson 1972, 155). The resident commissioners were not bound by the advice of the council, but, in practice, draft legislation and the budget were presented and the council's views on these were often taken into account (Kalkoa quoted in Jackson 1972, 155). At the previously mentioned meeting, an influential French businessman "stressed the necessity for a proper census as the basis for future planning" (Townsend 1964). This was the local manager of the Compagnie Français des Phosphates de L'Océanie, a company that was exporting manganese from a mine at Forari, on Efate Island, one of the first examples of mineral extraction of this kind in Vanuatu.[6] This endorsement was reportedly finally enough for the French authorities to agree to support a census. Thus, the census, connected with settlers' need for a wage labour force that was associated with the extraction of resources and changing land use, was set in motion.

During the planning stages, McArthur and the British and French administrators needed to agree on relevant population-related questions and categorizations. McArthur suggested asking the following:

1 Name
2 Relationship to the head of the household
3 Sex
4 Age
5 Marital status
6 For each woman aged 15 years or more
   a) The number of children she has ever born
   b) The number of children still living
7 Ethnic origin
8 Place of Birth: island of birth for the New Hebrideans, country of birth for others
9 Religion
10 If in paid employment, the kind of work [meaning employer] in which engaged (McArthur 1965)

These questions were discussed at a meeting of district agents (DAs) in Port Vila. All were approved except for question 10, to which the DAs made minor changes. Specifically, they indicated "that the question might more usefully refer to occupation rather than employer" (District

Figure 3.4. Front cover of census report
Source: McArthur and Yaxley (1968). Photograph courtesy of the author.

Agents 1965). After McArthur asked for more particulars about occupation, she settled on the categories shown on the Household Schedule card presented earlier in figure 3.2.

Training Ni-Vanuatu

Census cards needed to be prepared in three languages (French, English, and Bislama) and distributed to over eighty islands. As mentioned, it was not a census based on self-ascription; rather, it was carried out by enumerators, mainly local schoolteachers, seen above in figure 3.4. These enumerators required training to understand the census-taking process and recognize the categorizations, according to the requirements of Norma McArthur and the British colonial administrator John Yaxley.

After completing a pilot study in the Port Vila area, the team started to undertake the full-scale census across the archipelago. I can imagine that Ni-Vanuatu teachers on various islands, assured that they would be home again by Christmas, would have been intrigued by the opportunity to go to Port Vila for enumeration training during the December vacation. They may have received a letter based on the following template, or perhaps they received verbal invitations from the DA if their school was in proximity to the district offices:

> British Residency, Vila, New Hebrides
> 22nd October, 1966
>
> Dear ________________,
>
> As you may know the Condominium has decided to have a census in the New Hebrides in May next year. Some of you helped with the Census of Vila last year and already know what a census is and how it is organized. We are going to ask everyone who lives in the New Hebrides some questions about themselves, for example their name, their age, whether they are married, how many children they have, where they were born, what their nationality and race is, their religion and what their job is. The government wants to find out how many people live in the New Hebrides. And, more important, it wants to find out how many people there will be in the New Hebrides in 5 years or 10 years or 15 years' time so that it can plan new schools, hospitals, roads and train people as teachers, doctors and in other skilled trades and professions.
>
> We need about 180 people to go round the villages and get the answers to the various questions mentioned above. This is important work and we hope that you will be able to help either in your home area or in the area where you teach. The census has been planned to take place during the school holidays in May so that you will be free from your school duties. The holidays will be for three weeks so that you will have plenty of time to do the work and have a holiday. You will be paid about $2.50 for each day that you work.
>
> You should already have received from the Education section details of a Refresher course at Kawenu College in December. During this course, you will have some instructions in your duties as a census enumerator. The work is not difficult and we think you will enjoy it.
>
> We shall arrange transport for you to come to Vila and get you home for Christmas. You will be told what these travel arrangements are in the near future.
>
> We look forward to seeing you in Vila in December.
>
> J.F. Yaxley, British Director of Census. (Yaxley 1966)

The training consisted of lectures, practice sessions devoted to filling in schedules for an imaginary household, a trial census, and a period of revision during which a copy of the "Instructions to Enumerators" was given to each enumerator (McArthur and Yaxley 1968, 72–3). There were two training courses held in Port Vila (one before Christmas and one after) and one on Santo during the Easter vacation. Arranging ground and sea transportation for some eighty-five teachers to come to Kawenu College in Port Vila was an extraordinary feat. Those in the Vila area had likely worked on the pilot study. The mornings were taken up with lectures, and during two early evening sessions, would-be enumerators practised census taking in villages near Vila, Pango, Mele, and Maat. Chiefs Andy Riaman (Pango), Peter Poilapa (Mele), and Enoch (Maat) were consulted and thanked for their cooperation.

At the training on how to use the Household Schedule and record the information properly, the enumerators were once again informed why the census was important. The following excerpt is from a training handbook distributed to enumerators and field supervisors:

Some Reasons for the Census

1 This is the first general census of the New Hebrides. It will tell us how many people live here. It will also give us some idea of how quickly the population is growing.
2 It will tell us how many children are ready for school in 1967 and in the following years so that [the] Government and missions will know which areas most need schools and teachers.
3 It will show us which places have many deaths from sickness so that the medical authorities can pay particular attention to these areas.
4 It will show how many people move from their home islands to live on other islands.
5 It will tell us what kind of work people do.

The information which is obtained from the Census will help the medical, education and agricultural authorities and other central Government Departments together with Local Councils, to plan more and better services to help the people of the New Hebrides Condominium. It will help planning for the future.

DO NOT FORGET. The success of the Census depends on you because you are the person asking the questions and filling in the household schedules. If you are lazy or not careful the census will be a failure. So please do your best. (NHBS 1967b)

In the way that the census is justified, Ni-Vanuatu were enrolled in state politics and a moralized work ethic. Each enumerator was assigned an area with approximately a hundred households or four to five hundred people. McArthur and Yaxley (1968, 72) reported that

> revision courses were held at Vila, Santo, Lakatoro (Malekula), and Tanna for those enumerators who could travel to these centres without difficulty. Those who were unable to attend were visited in their villages or at their schools by one of three census training teams which were formed to travel throughout the Group between January and April for this purpose. These visits gave the census teams an opportunity for meetings with the enumerators and their Field Supervisors, to solve any outstanding problems and to distribute the census materials to Field Supervisors.

Representatives of minority groups (e.g., Europeans and Asians) enumerated their own groups. Each field supervisor – these were mainly European missionaries and settlers – was placed in charge of a group of enumerators. Training for the enumerators was also provided in French. One such training took place in Tongoa on 3–4 March 1967. The participants did a practice census in the village of Panita (NHBS 1967a). In the way that she went about the census, McArthur followed her own advice as laid out in her handbook.

The field supervisors needed training too. Marnie Anderson, an Australian assistant to Yaxley, visited the field supervisors in as many areas as possible to explain their duties. She would forward them the Household Schedule, pens, plastic covers, and maps once they arrived from Sydney or Noumea (Yaxley 1967). While making these tours, she found that the census song that had been broadcast on the radio had been heard by a wide audience. The coverage was so good, she said, that the French priest Père du Rumain reported that people were coming to see him to consult the parish register to determine their ages (Wallmark 1967).

Household: Documenting Social Relations around Shared Food

Figuring out how to classify a household was also no simple matter, as the occupants of a given building might shift from night to night and households might not overlap with family groupings in a Western or biological sense of the term. After a discussion with representatives from major island groups, the census designers concluded that a household would be defined as a group of people who usually ate together and prepared their food in the same kitchen (McArthur and Yaxley 1968, 68), thereby creating a countable unit connected to an economic subsistence pattern. Even still, it was hard to isolate households, especially in areas where daily activities were "communal" rather than "individual," according to the census team. The census made distinctions between private and non-private households. Non-private households were prisons, barracks, ships, and hotels. Enumerators were instructed that married people and their offspring should be entered into the census forms in an identifiable way within the household. The married couple, with or without children, was the basic family unit in this respect, and "by definition, non-private households contained no family units" (68). Membership within a household was categorized in terms of individuals' relationships to the "head of household," who was a man. All others were categorized in relation to this man as per the design of the census card. Within the tables published in the census report, households were depicted as units in terms of size (total number of people and number of families) and location, but neither individual nor household incomes or wealth were recorded.

There was to be one census card per household. The materiality of the card compelled the enumerator to pay close attention to this grouping categorized according to who ate together and how they were associated with the male head of household. These were, practically speaking, relationships that could be written down, were easily talked about, and thereby made socially apparent. Through the enumeration of social relations revolving around eating and cooking (which women were central to organizing), men were given a central position. The household would ultimately be analysed within the chapter in the census report devoted to "Households and Family Units." The average household size was 5.06 people, while the average family unit was 4.57 people (McArthur and Yaxley 1968, 70). On the household, McArthur and Yaxley (1968, 68) made the following observation:

> [The] usual indexes of overcrowding of 2 or more families in a house are not valid. However even in the New Hebridean component, especially in areas where the extended family household is giving way to a more European pattern of family living, the co-existence of several family units in a large proportion of all households with many of them containing more than 2 family units, may indicate pressure on housing availability.

Fertility: Documenting Biological Relations between Women and Children

Like demographers and colonial planners at that time, McArthur and Yaxley were interested in fertility rates (measured in terms of the number of births per woman). In the 1967 census, three (out of twenty-five) tables are devoted to establishing the relationship between women, their ages, and their number of children. In her handbook, as well as in the 1967 census, McArthur includes long passages about why women's life course matters for fertility measures (McArthur 1961). There is no mention of men's fertility, which could actually be much higher as they could have more than one child in nine months. This standard demographic practice, following a logic that determined what was practical to measure, contributed to making sex-specific medical fertility control self-evident.

When Norma McArthur headed up the census in the New Hebrides, she was already a leading demographer of the Pacific Islands and had her own particular research interests. Understanding population structure was central to her aim of providing governments with solid baseline data so that they could determine future population trends (Campbell 2006, 99). The mechanics of "population trends were also susceptible to the direct effects of migration either inward or outwards, but the most significant variable was the size and fertility of the young adult female population" (Campbell 2006, 101). McArthur was also interested in demographic structure because she aimed to show that "an epidemic will not cause sustained population decline unless it affects the age or sex structure of the population in specific ways. Thus, it is less important how many people were killed, than which ones were" (Campbell 2006, 103).

This emphasis on understanding population structure meant following changes in fertility over time. Demographers measure and project fertility over time through female bodies and their life cycles, wherein fertility has measurable starting and ending points. Ideally, demographers measure fertility rates (births per woman over her lifetime) and birth rates (total number of live births in a population per year). In Vanuatu, calculating birth rates was not feasible because birth registrations were neither compulsory nor universally possible. This led to a methodological emphasis on asking women about their pregnancy and birth histories. In order to generate data that would serve to answer McArthur's questions about population structure and fertility rates, irrespective of marital status, all women aged fifteen and over were asked three interconnected questions: "(i) the total number of children borne by them, (ii) the number of these still living and (iii) the woman's age when her first child was born" (McArthur and Yaxley 1968, 44; see figure 3.5). As previously noted, the census cards were organized so as to enable the collection of data based on relations of social reproduction (food preparation and eating) that organized households around the male head of household. This design meant that enumerators had to be instructed

NAME OF ISLAND NOM DE L'ILE	1 & 2
ENUMERATORS AREA CODE CODE DE LA ZONE DU RECENSEMENT	3 & 4
NAME OF VILLAGE OR REGION NOM DU VILLAGE OU DE LA REGION	5 & 6
SCHEDULE NUMBER NUMERO DE L'ANNEXE	7, 8 & 9

CONDOMI
CONDOMIN
CENSU

This **HOUS**
completed by an
AREA for which I

Cette **ANN**
tribuée ou rempli
dans LA ZONE dc

1	2		3	4		5			6	
NAME **NOM**	**RELATION-SHIP TO HEAD OF HOUSE-HOLD** e.g wife son, visitor, etc.	**PARENTÉ AVEC LE CHEF DU FOYER** p. ex. épouse, fils, visiteur, etc.	SEX SEXE M or/ou F	AGE LAST BIRTHDAY if under 1 year write "0"	AGE DERNIER ANNI-VERSAIRE si moins d'un an, écrivez "0"	FOR FEMALES AGED 15 YEARS OR OVER / POUR LES FEMMES AGÉES DE 15 ANS OU DE PLUS			MARITAL STATUS NM = Never Married M = Married W = Widowed D = Divorced	SIT / M JM Jar Ma M Ma V Ve (Ve D Div
						Total number of children/ Nombre total d'enfants		Age when first child born/âge lors de la naissance du premier enfant		
						Born/ Nés	Still living/ Vivants			
12			13	14	15	16 17	18 19	20 21		

Figure 3.5. Census schedule: Note questions 2 and 5
Source: McArthur and Yaxley (1968). Photograph courtesy of the author.

to identify any children present in the household but not borne by the woman answering the questions, and to enquire about other biological children who may have been adopted into other households or were living elsewhere. So, while relationships within a household were enumerated in relation to a male head of household, thereby privileging social relations that revolve around men, information collected about children in relation to their biological mothers served to link women with the biological aspects of reproduction.

Subsistence: Making Socio-Economic Aspects of Reproduction Visible

While partial censuses had been undertaken sporadically for decades, the methods consisted mainly of counting people by village and island. The most complex calculation was that of the sex ratio, as discussed in chapter 1. The 1967 census, which built on the Port Vila pilot project, was the first time that people's relationship to a cash economy was formulated across the islands. McArthur was completely aware that the census imposed categorizations that simplified a more complicated social reality. But codable definitions of livelihoods had to be chosen that would facilitate enumerators' work in the field and allow them to yield relevant data that could then be properly analysed. Ultimately, two questions were selected for initiating this topic: "What work do you do?" and "Who do you work for and where?" This selection was intended to accommodate the fact that, while money was used throughout the archipelago, most Ni-Vanuatu, it was thought, were not dependent on cash incomes but, rather, were situated somewhere along a continuum between "pure subsistence" and "the monetary economy." The two questions on occupation and industry were phrased as simply as possible so that the answers could fit with the industry classification. Enumerators were instructed to ensure that the replies they recorded included sufficient information to enable the census staff to accurately classify each person according to his or her occupation and industry (McArthur and Yaxley 1968, 57).

Because the exact nature of respondents' "work" was often ambiguous, the census team provided instructions for how enumerators should fill in the Household Schedule, and these were tested during the pilot study in Vila, where the wage labour categories were more relevant than they were in many other places in the archipelago. In Bislama, enumerators were given the following instructions on "occupation": "Work bilong man. Olsem police, driver, nurse, man i work long ship, housegirl, school picannini, man he work long wharf. Sipose woman where i stap look out family bilong hem no more, writim housewife. Sipose man where i no gat work, writim no." (The work the person does, such as police, driver, nurse, boatswain, domestic servant, school child, dock worker. If a women just takes care of her family, write "housewife." If a man does not have work, write "no.") (Brookfield, Brown, and Anderson 1966). The instructions on "employer or school" were as follows: "Nem bilong ples more man where i work more school long hem. Olsem B.P. Wharf, British PMD, Ecole Publique, Missus Hulot. Sipose man i stap nothing, i no writim ene something long place ia." (Name of place and person and where they work or go to school. Like B.P. Wharf, British PMD, the French public school, Mrs. Hulot. If the person is doing nothing, do not write anything in that place.) In English, the following instructions about work were provided:

> "Occupation." The nature of the work should be described as precisely as possible, for example: Administrative officer, shop proprietor, shop assistant clerk, mechanic, nurse, plantation manager, boatswain, seaman, policeman, stevedore, general labourer, taxi driver, teacher, typist, housegirl, student, retired. If there is no employer leave the space blank. "Employer." Please be specific. For example: Burns Philip Wharf, S.M.E.T, Bureau de la Residence de France, British Works and Stores, Ecole Publique, Hotel Vate, "Tutuba,"

> C.F.N.H. Store, Kawenu College, Mme. Hulot. If there is no employer, or school, leave the space blank. (Brookfield, Brown, and Anderson 1966)

On the census form itself, the enumerator would proceed to a different set of questions concerning occupations, depending on the answer received to the question "What work do you do?" (subsistence sector) or "Who do you work for and where" (monetary sector). Given the varying degrees of participation in the monetary economy, even in the "subsistence" sector, McArthur had the enumerators further categorize livelihoods in the subsistence sector (implicitly in villages) as "villager, no cash crop" and "villager, gardener and copra maker." McArthur and Yaxley (1968, 57) admitted these were arbitrary and subjective. The copra (processed coconut) production activities could also be classified differently as well. Enumerators were told to distinguish between those who made their own copra, either for their own purposes or to fulfil communal obligations, and those who cut and prepared copra for sale (58). The former were coded under "subsistence and village agriculture," and the latter were described as copra cutters and were classified under "plantation coconut growing." The wives of plantation workers who accompanied their husbands were normally coded under "home duties," unless there was evidence on the census schedule that they were working alongside their husbands as copra cutters. Finally,

> Enumerators were instructed to use the blanket classification "Gardener and Copra Maker" to include all rural activities within the subsistence sector such as basket and matmaking, canoe building, etc., in addition to gardening. The classification did not include such occupations as full time storekeeper or village carpenter, because such persons were more outside than inside the framework of the subsistence sector. (58)

Beyond the categories "villager, no cash crop" and "village gardener and copra maker," McArthur and Yaxley did not develop further classifications for the subsistence economy because, as she wrote, "in view of the nature of subsistence and village agriculture there can be no division into grade or type of occupation (i.e., status) as there is in those industries within the monetary economy" (1968, 57). The categories she assigned to the monetary economy (and note their hierarchical organization) were "(i) professional, management and executive; (ii) supervisory and clerical; (iii) skilled and semi-skilled (combined because the low level of industrial development and in particular the absence of formal trade training make it impossible to distinguish the two groups with any degree of accuracy); and (iv) all other" (57). McArthur and

Yaxley categorized participation in the monetary economy within a hierarchy of ranked occupations. This scheme amounted to a correlation between occupations associated with the monetary economy being linked with people's identity categorizations. So the broad majority of the population – Ni-Vanuatu – were associated with the allegedly uncomplicated subsistence economy without internal hierarchies (see tables 3.1 and 3.2 below), while a much smaller minority was associated with economic activities that could be described as either more or less complicated. McArthur and Yaxley's broad definition of subsistence and monetary economic activities meant that men's and women's participation could be accounted for in economic activities. And their respective levels of activity were quite similar: 86.5 per cent of men and 81.9 per cent of women were "economically active." The various activities associated with subsistence and village agriculture occupied respectively 69 per cent and 91 per cent of the "economically active" men and women within the New Hebridean population at the time of the census (McArthur and Yaxley 1968, 60).

These figures were remarkable! Prior to World War II, and especially in the 1910s and 1920s, researchers saw Pacific Islanders as a dying "race," biologically vulnerable and culturally ill-equipped for the modern world. Norma McArthur's censuses, by contrast, showed that Pacific Islanders were strong, more numerous than expected, able to cope, and even prosper, with modernity.[7] Importantly, they showed not only that Indigenous people were not dying out, they were in fact economically active.

Also remarkable, and yet taken as self-evident, was the fact that so many people had access to land, enabling them to live lives as subsistence gardeners, thus preventing wide-scale dependency on wage labour. Changes in this respect were, however, on the horizon. McArthur and Yaxley made the following remark relating to their finding that young men were working in wage labour in larger numbers than expected: "The younger New Hebridean males are already drifting away from subsistence and village agriculture towards more sophisticated occupations" (1968, vii). Recall, too, that the French businessman's desire for population numbers for the expansion of wage labour was a key motivator that finally prompted the implementation of the census.

Throughout the correspondence in the census files, and in the final published report, administrators' and demographers' emphasis remained consistently on planning for the future by providing medical and educational opportunities. The census rendered the labour and biological reproduction of men and women visible. It highlighted the significance of men's status as heads of households and the increase in

Table 3.1. Proportions (per 1,000) of the economically active adult males and females in each component populations who were engaged to each major group of industries

	New Hebridean		European		Chinese		Vietnamese		Other Melanesian		Polynesian and Micronesian		Part-European		All Other Mixed and All Others		All Components	
Industry Group	M	F	M	F	M	F	M	F	M	F	M	F	M	F	M	F	M	F
Subsistence Agriculture	48	54	0	0	0	0	0	0	0	23	0	0	0	0	0	0	43	52
Village Agriculture	646	860	0	0	0	0	0	0	69	273	9	0	58	120	24	231	585	826
Plantation Agriculture Including Livestock Production	99	21	149	22	0	0	83	18	69	0	427	71	204	46	12	0	108	21
Other Primary Industry	20	0	42	10	454	0	92	18	41	0	201	71	39	9	443	38	33	1
Secondary Industry – Manufacturing	13	2	54	35	56	130	92	161	62	0	140	14	97	38	286	115	22	3
Secondary Industry – Construction	28	0	106	22	63	0	267	18	172	0	107	14	150	0	109	0	37	1
Secondary Industry – Commerce	25	1	123	113	364	784	133	410	41	0	26	86	170	194	48	193	33	8
Secondary Industry – Transport and Communications	36	1	91	28	63	0	117	18	207	0	46	14	160	56	36	0	42	2
Secondary Industry – Other Services	12	38	43	35	0	0	91	89	34	386	15	329	44	194	12	231	13	42
Administration and Government	12	0	105	138	0	0	0	36	0	23	7	43	19	56	12	0	15	4

(Continued)

Table 3.1. Continued

Industry Group	New Hebridean		European		Chinese		Vietnamese		Other Melanesian		Polynesian and Micronesian		Part-European		All Other Mixed and All Others		All Components	
	M	F	M	F	M	F	M	F	M	F	M	F	M	F	M	F	M	F
Professional and Allied Services	56	21	275	525	0	43	58	161	276	295	7	71	39	213	12	154	63	34
Other Economic Activities	5	2	12	72	0	43	67	71	28	0	15	287	19	74	6	38	6	6
Population Economically Active	855	845	942	553	973	500	889	475	980	494	987	261	932	500	994	500	865	819
Population (Nos.) Economically Active	17,529	15,070	650	318	143	23	120	56	145	44	458	70	206	108	165	26	19,418	15,715

Source: McArthur and Yaxley (1968, 61).

Table 3.2. Proportions (per 1,000) of the economically active adult males and females in each group of industries from each component population

Component Population	Subsistence (s) and Village Agriculture (v)		Plantation Agriculture including Livestock Production	Other Primary Industry	Secondary Industry – Manufacturing	Secondary Industry – Construction	Secondary Industry – Commerce	Secondary Industry – Transport and Communications	Secondary Industry – Other Services	Administration and Government	Professional and Allied Services	Other Economic Activities
New Hebridean	s	v										
Males	503	465	717	542	480	670	570	761	214	596	599	428
Females	496	532	129	14	48	3	18	13	627	17	179	144
European												
Males	0		40	41	73	93	104	71	31	197	102	38
Females	0		3	4	23	10	47	11	12	128	95	111
Chinese												
Males	0		0	99	17	12	67	11	0	0	0	0
Females	0		0	0	6	0	23	0	0	0	1	5
Vietnamese												
Males	0		4	17	23	43	21	17	12	0	4	38
Females	0		0	1	19	1	30	1	5	6	5	19
Other Melanesian	s	v										
Males	0	0	4	9	19	34	8	36	5	0	23	19
Females	1	1	0	0	0	0	0	0	19	3	7	0
Polynesian and Micronesian												
Males	0		81	140	134	66	16	25	8	9	2	34
Females	0		2	8	3	1	8	1	25	9	3	96

(Continued)

Table 3.2. Continued

Component Population	Subsistence (s) and Village Agriculture (v)		Plantation Agriculture including Livestock Production	Other Primary Industry	Secondary Industry – Manufacturing	Secondary Industry – Construction	Secondary Industry – Commerce	Secondary Industry – Transport and Communications	Secondary Industry – Other Services	Administration and Government	Professional and Allied Services	Other Economic Activities
Part-European												
Males	1		17	12	42	42	45	39	10	12	4	19
Females	1		2	1	8	0	27	7	23	17	13	39
All Other Mixed and All Others												
Males	0		1	111	99	25	10	7	2	6	1	5
Females	0		0	1	6	0	6	0	7	0	2	5
All Components	s	v										
Males	834	11,357	2,092	639	423	729	649	810	257	282	1,225	121
Females	824	12,984	329	20	54	11	123	28	655	62	538	87

Source: McArthur and Yaxley (1968, 62).

men's wage labour. "Subsistence" is a term that obscures the horticultural knowledge and social networks contained within social and biological reproduction. What this term flattens are Ni-Vanuatu skills in animal husbandry, of pigs in particular, and growing plants, like yams and taro, that are important for their place in exchange networks. Having access to and being able to produce such plants and animals was an important means of gaining status and participating in significant life course events like funerals, marriages, or births. In the census categories, "subsistence" became a placeholder for social reproduction and made complex networks visible in a knowledge category implicated in state plans.

Those gardening activities glossed as "communal obligations" in the census involve moral relations and duties. They could have been part of everyday survival or large-scale rituals. As an example, Jean Mitchell writes eloquently about the connection between gardening and a significant ritual called *nieri* on Tanna, in southern Vanuatu. This event, which involves the exchange of food tubers between partners in several villages, "assert[s] the material and spiritual primacy of gardens and the exchange of garden food" (Mitchell 2021, 433). Furthermore, growing food for this ritual is part of reproduction, as the "exchange is charged with remaking persons and recognising the various persons that render growing and becoming, transformation and regeneration possible" (434). Mitchell's research also shows that through Tannese gardening practices, humans and non-human beings facilitate both spiritual and material growth. It is recognized that gardening throughout the south-western Pacific Islands includes technologies and disciplines that "evince gardeners' knowledge of right action" (Battaglia 2017, 280), because plants (yam and taro in particular) are affected by "human feelings, substances and transgressions" (Mitchell 2021, 440). In this way, such gardens "extend the moral and social relations of humans to other non humans" (441). Finally, gardening is about maintaining connections with human and non-human ancestors. One of Mitchell's interlocutors, for example, lamented the loss of ancestral crops of yam and taro in a 2015 category 5 cyclone, saying, "We should have taken better care" (442). Mitchell explains that this was not only a loss of heritage tubers, but also a loss of relations with her interlocutor's human ancestors.

The figure of "subsistence" collapses all of these intricacies, and in particular the spiritual and material importance of land access. These relations produced and reproduced through production and exchange are opportunities for persons, food, and shelter to be produced and reproduced, and for right relations to be demonstrated to plants, humans, and ancestors, all in the service of human growth and regeneration. In referring to "subsistence" as a moral figure, I am attempting to call attention to these dimensions, which were occluded in the census takers' quantifications and plans in an effort to enable participation in the wage economy that was expected to replace the subsistence economy.

The ways in which censuses produce and reproduce (rather than merely enumerate) particular aspects of identity, like race or ethnicity, have been of interest to many scholars (e.g., Rodríguez-Muñiz 2017). In the training materials provided to the Ni-Vanuatu enumerators, there was no discussion of how to recognize the desired categorization of "ethnic origin"; it was simply taken as self-evident. The category of "race" was an implicit part of this categorization, as we see in the brief instructions on "ethnic origin":

> When there is any doubt regarding a person's racial origin, the answer to be written is the race to which the person himself thinks he belongs. Races other than those listed at the top of the column may be used if they do not seem to give the right answer. Any person who is not a full European should be called "part-European" if he has any European blood at all. (NHBS 1967b)

Practically speaking, enumerators who circulated among Ni-Vanuatu villages would primarily encounter people living the "native way of life," a colonial designation that had specific legal implications. "Half caste" individuals with one British parent could gain legal status (notably the right to own land individually, as opposed to collectively), or could lose their provisional citizenship status, depending on which "way of life" they were leading (this was a category on official documents; see Widmer 2013).[8] Therefore, a person with a British parent would be given provisional British citizenship status if they lived as a European (in town, in a European-style house, speaking English or French), and they could lose it as well if they failed to abide by these expectations. The census category "man New Hebrides," then, was virtually synonymous with living in a village and "the native way of life."

Reproductive Silos: Land, Birth, Education, and Labour

Though one could not tell from the census categories or the overriding planning goals by which such an endeavour was justified, the 1960s were a busy time for the French and British Residencies in the areas of land use and ownership policies. Such changes would have foretold enormous planning implications for instrumental reasons alone.

Land access was the basis of widespread participation in the subsistence economy. In the colonial paper trails and administrative meetings, other than those pertaining to the census preparation, changes in how land would be used figured prominently among the concerns of the colonial authorities. Ni-Vanuatu worries were also evident in this respect. For example, Dr. John Kalsakau from the island of Ifira in Port Vila Harbour (and a member of one of the region's most powerful families), along with other Ni-Vanuatu on the New Hebrides Advisory Council, wanted to form a land commission to look into land registration. Dr. Kalsakau was one of the first Ni-Vanuatu to train as a physician, and he had been advocating to the Condominium for improved health conditions for Ni-Vanuatu since 1941 (Widmer 2010).

Despite its absence in the census correspondence, the question of land was clearly on administrators' minds during this time. In 1960, BRC Rennie once again initiated conversations with FRC Delauney on the "position of native custom relating to land owned by New Hebrideans, in the light of changing conditions, and the land law of the Condominium and the increasing interest in the production of cash crops" (Van Trease 1987, 94). In 1964, the BRC invited Rowton Simpson, the land tenure Adviser to the secretary of state for Commonwealth affairs, to travel to Vanuatu from London (Van Trease 1987, 98). Simpson stressed the importance of sorting out the land tenure question because of increasing interference with survey teams and the feeling among people that their ancestors had been misled and cheated. In a subsequent report, Simpson recommended that unused land held by foreign individuals and companies be made available for Ni-Vanuatu to use, and that measures to "compel the owner of land which is unused or misused to make proper use of it" be adopted (Van Trease 1987, 99). He also called for improvements to the registration system so as to enable Ni-Vanuatu to register land if they wanted to, and for initiatives that would "assist them in the transition from subsistence to commercial agriculture" (Van Trease 1987, 99). His recommendations about transferring unused settlers' land back to Ni-Vanuatu were, however, omitted in his report submitted to the Advisory Council in 1965 (Van Trease 1987, 99).

In the 1960s, and especially during the latter part of that decade, the pace of capitalist economic development intensified in Vanuatu, as did Ni-Vanuatu protests over land access and/or compensation for land. Ni-Vanuatu would see changes in economic production manifested, for example, in changes to land use and in the fences erected around settlers' cattle pastures. Ni-Vanuatu recognized in these changes the effects of European land claims being registered at the Joint Court that had previously not been fully apparent.[9] Ni-Vanuatu resistance to this process of land registration took different forms. One domain was on the Advisory Council. In December 1965, Dr. Philip Ilo, claiming to speak on behalf of all Ni-Vanuatu, stated that all the Indigenous council members regarded the decision of the Joint Court as "improper" (Pacific Islands Monthly 1967, 32). Another example is that of the counter land claims made by Fila Islanders to the Joint Court pertaining to the construction of the wharf in Vila Harbour. Led by Chief Graham Kalsakau, these Islanders tried to use land registration to secure their own claims, as they resented the fact that profits were being made on land they believed was rightfully theirs (Van Trease 1987, 188).

Another dimension of resistance to land use and profits was evident on the island of Espiritu Santo, Vanuatu's largest island. On the southeast corner of Santo, the French colonial company Société Française des Nouvelles-Hébrides (SFNH)[10] had a holding of some thirteen thousand hectares. French planters who had purchased or leased land from the SFNH had begun enclosing this land for their beef cattle plantations, the meat from which would be sold in Noumea, New Caledonia. Previous cash crops of cocoa and copra had required far less land than was the case with beef production (Jackson 1972, 158). Around the same time, Ni-Vanuatu living further inland on Santo had been moving down towards the coast and occupying land for subsistence purposes and in order to be close to sources of wage labour and schools for their children.

As reported in the *Pacific Islands Monthly* in May 1967, Chief Buluk wrote to the SPC, the regional development organization, to publicize a case in which a European had landed a hundred rolls of wire in the Big Bay area. He asked the SPC to "please stop these planters not to do this thing anymore" because "all the dark bush" in the area belonged to him and "coming native children." He claimed that he had reported this activity ten years earlier to both the French and British DAs (*Pacific Islands Monthly* 1967, 31). The *Pacific Islands Monthly* writer described the situation as follows: "Although there has always been an undercurrent of discontent among the New Hebrideans over past land transactions with Europeans, it is only in recent years, as the population in the Group has been increasing, that the New Hebrideans have become outspoken on the subject" (32). While the census confirmed that the population was indeed increasing, this was not the only relevant concern. The rise in disputes also came at a time when land use was changing and wage labour was expanding, with a growing number of Ni-Vanuatu working for settlers.

Chief Buluk, the leader of a group of about fifty Ni-Vanuatu, began working with Jimmy Stephens, the grandson of a Tongan woman and English man. Stephens encouraged the group to remove the fences that French plantation owners had erected to fence in their cattle. After Jimmy Stephens moved to the area, it became the base for an anti-colonial movement called Nagriamel.[11] He would go on to accuse the Ni-Vanuatu members on the Advisory Council of being elites who were mainly interested in collaborating with colonial authorities. By 1969, the Nagriamel movement reportedly had ten thousand supporters, most of whom were located in the northern and central areas of the archipelago. Over the coming years, Buluk and Stephens continued to organize together. In January 1966, they held a public meeting on Santo that was attended by over six hundred men, women, and children (Van Trease 1987, 139). Those present participated in the ongoing declarations of the manifestos called "Act of Dark Bush Land," claiming all land that had not yet been developed by Europeans (like the cattle grazing grounds). At the peak of the movement, they succeeded in occupying sixteen hundred hectares through direct action and squatting.

According to Van Trease (1987), there followed years in which there was no cooperation between Britain and France on land issues. The British authorities saw independence as a possibility for Vanuatu and believed that the unused land owned by foreigners would be a source of conflict when they left. On the other hand, France was not interested in independence or in opening up the question of unused land.

Conclusion

As Sethy Regenvanu's words in the epigraph to this chapter indicate, for Ni-Vanuatu, land is the moral heart of reproduction in both a biological and a social sense. Subsistence, when mobilized as an economic population category, is thus a placeholder that makes a complex Indigenous system of knowledge and practices into a development category that is outside of GDP but which, as Mitchell (2011a) argues, can successfully support populations. The category of "subsistence" renders Ni-Vanuatu economically active but in need of preparation for wage labour. Quantifying the number of people who participated in subsistence was essential for planning purposes, and this was the promised outcome of the census. This was a kind of planning that would allow people to attend school and access medical care provided by a state. It was a kind of planning in which Ni-Vanuatu participation in wage labour (generally for settlers or colonial administrations) was seen as inevitable – a goal of modernization and development. It enrolled Ni-Vanuatu in a kind of politics whereby work is made visible as "subsistence" and wage labour is understood as a form of modernity into which they were conscripted (Asad 1992; Scott 2004). Thus, the resulting plans signified well-intentioned concerns for a presumed future in which Ni-Vanuatu would become participants in a cash economy, rather than participants in the political structures and subsistence patterns of their ancestors. There is no mention anywhere in the census of land tenure, despite the central importance of subsistence for survival. Neither is there any mention of the importance of land access, let alone land ownership, as a way of ensuring the continuation of subsistence livelihoods.

In his discussion of the workings of colonial power, Scott argues that "the political problem of modern colonial power was therefore not merely to contain resistance and encourage accommodation but to seek to ensure that both could only be defined in relation to the categories and structures of modern political rationalities" (1995, 214). During the period under study, colonial power in the New Hebrides entailed rendering birth rates and population growth processes in ways that would allow them to be measured and intervened on separately from land ownership and land use policies. The results of the census showed that the population was growing and needed to be trained for wage labour. There was concern that population growth might put pressure on land. The concerns about land use – more land was being used by settlers to grow food for export – were made visible for governmental intervention separately from the census.

Unpacking the significance of statistical constructions of Indigenous peoples is vitally important, as these numbers are implicated in "demographic classification, state policies, indigenous response and indigenous identity" (Axelson and Sköld 2011, 12). In this chapter, I have shown how McArthur's censuses and demographic analysis presented the islands as places where Indigenous populations were vigorous and economically active. Ni-Vanuatu were made visible and portrayed as being in need of planning to support their inevitable transition to wage economies and state policies. The data-collection practices and categories applied in the census enrolled Ni-Vanuatu in a politics of visibility whereby Indigeneity was produced in relation to economic livelihoods and the monetary economy. The census, and the demographic projections it enabled for the purposes of state planning, thus enrolled Ni-Vanuatu in a form of state politics that separated education and medical planning and access to wage labour while making subsistence visible. The census rendered subsistence, a knowledge category of economic production and social reproduction, separately from land politics.

The census results showed that the Ni-Vanuatu population was larger than expected and emphatically not dying out. McArthur expected the population to be 65,000 people at the most, but the census made clear that it was in fact 76,582, of whom 92 per cent were

Ni-Vanuatu (McArthur and Yaxley 1968, vii). Further, the population was growing at a rate of 2.5 per cent and would likely double in thirty years (viii). The expansion of schools and programs for maternal health were therefore necessary, and these would in turn enable the growth of the wage labour force.

With the figure of "subsistence," then, the census made both biological reproduction and social reproduction visible in ways that would foster development agendas and the growth of wage labour; at the same time, they elided land politics from social and biological reproduction. It did so by emphasizing male-headed households who shared cooking facilities, making women's fertility responsible for birth rates and linking Indigeneity to subsistence but not to land access. I am not arguing here that these outcomes were the result of any bad intentions on the part of Norma McArthur and her team. Rather, I am suggesting that they are a result of the production, by way of the census, of the knowledge category of "subsistence," which collapsed the relationship between Ni-Vanuatu knowledge and work and the kind of planning the census was purported to do. Though the census reported that over 80 per cent of men and women were active in subsistence agriculture, what it did not do was make planning recommendations that would ensure access to land in the face of expanding agricultural and mining practices that would take land and labour away from those trying to maintain "subsistence" ways of life.

In focusing on colonial and development processes as they relate to the production of the figure of "subsistence," I hope to have shown in this chapter how quantification elided the importance of land and the collective relations that are central to Ni-Vanuatu socialities. Relationalities between people, land, and knowledge, which Ni-Vanuatu consider to be moral, underpinned social reproduction and made a robust way of life.

Chapter Four

"I Just Wanted to Be Invisible": "Young Mothers" from Global Discourse to Village Experience, 2010–2020

"Happy!" shouted the parade leader into his megaphone. "Children's Day!" responded the crowd enthusiastically while both participants and spectators waited on the Port Vila waterfront. After three or four rounds, the chants died down and top 40 music hits returned to the loudspeakers on the leading truck. Eventually, the brass band of the Vanuatu Mobile Force, a paramilitary police force, arrived on the back of a shiny Toyota pickup, clad in black berets and camouflage uniforms. They lined up into marching formation and, on the cue of the conductor, launched the International Children's Day parade with a rousing rendition of the well-known tune "Colonel Bogey March."

Along with several other families from Pango Village, my family and I had made the approximately five-kilometre bus trip into Port Vila on this school holiday to watch the floats staged on brand new pickup trucks or rusty flatbed vehicles. School children carried banners and proceeded on foot. The parade ambled down the main street of the capital. On many of the banners, the parade's participants – representing schools, church organizations, and NGOs – had created colourful renderings of the national slogans for the day: "Investing in the Nation's Future" and "Protecting Children and their Families." Many waved Vanuatu flags as they held their banners proclaiming education and health care as rights that were *stamba blong development long nesen ia, Vanuatu* (the foundation of development in this nation, Vanuatu). Towards the end of the parade, children dressed in ancestral clothing such as grass skirts and donning traditional face paints (everyone else wore knee-length shorts and t-shirts or *aelan dres*) sat on a float proclaiming, *Mama and papa, no salem graon blong yumi!* (Mom and dad, don't sell our land!) and *Foreign education imas joinem witim kastom education blong gudfala fiuja blong nesen ia* (Foreign education must join with Indigenous education for a good future for this nation).

The parade moved along the main street and up the hill to the Saralana stage, the main outdoor venue for public events in Port Vila. As is typically the case at many similar public events on this stage, the people working at the relevant government ministries were thanked for their support, and they in turn gave polite speeches followed by perfunctory applause. During his speech, a senior program manager at the Department of Health rose and, referencing World Population Day, which had taken place a few days before under the slogan "Everyone Counts," discussed the importance of good census information in order to ensure that government programs reached all demographic groups in Vanuatu. The November 2009 census had only recently been conducted. In his Children's Day speech, the civil servant said the census would help secure a better future for the largest demographic group in Vanuatu: children under the age of fifteen.

Though this address about population growth and census categories did not rank among the highlights of the day's events, that a civil servant would make a public speech like this on Children's Day registers the profoundly significant role demographic knowledge plays in shaping public life in Vanuatu. Here, the public circulation of demographic knowledge connects the agendas of global international organizations, the ambitions of the national government, and the social needs of the village, who all face the challenge of providing for a growing population. Those who were present at this festive event participated in the circulation of a public discourse concerning reproduction in urban Vanuatu. It is a discourse that reflects the powerful influence of the population and development modelling of international organizations. This discourse, an assemblage that includes research methods, demographic indicators, narrative policy reports, and public celebration days, connects the future of a population to health and education as both "investments" and "rights." The control of both individual and population-level fertility takes place on the terrain of finances and individual empowerment. The force of such technocratic, institutional discourses sits – sometimes comfortably, sometimes awkwardly – alongside slogans reflecting collective awareness that Ni-Vanuatu children's future interests are under threat from land sales, and that familiarity with the archipelago's traditions is as important as formal education.

Another day, while bringing my older daughter home from the preschool in Pango Village, I stopped to greet and exchange news with a woman with adult children. She invited me to a fundraiser for a local church. She put down her baskets and told me all about her many church and family activities. She had been working at a pastor's garden, assisting with the fundraiser, and taking care of children. Like many women of her generation, she had worked hard her entire life.

"I work for the community," she said. "I always think about others, not like those young mothers. When I don't help out, I feel guilty. When you give, then you can receive. Young mothers, they just think about themselves, and kava and Tusker" (a brand of beer). These were common sentiments expressed about "young mothers"[1] and their perceived unwillingness to recognize the expectations and collective futures that their mothers or grandmothers feel they worked towards. Women of this generation work very hard for the community, and yet so, too, do "young mothers." In this chapter, I aim to contextualize such comments about "young mothers" as they point to values and practices in Pango regarding the place of young women in reproduction and their connection to securing good collective futures for their families.

As the Children's Day parade banners show, "investment" is a dominant trope through which the connections between reproduction and a good future are made in demographic measurements and population discourses. This economization of reproduction, instantiated in, for example, maternal health metrics (Erikson 2016) and maternal health infographics (MacDonald 2019), began in the overlapping Cold War and postcolonial history of demographic research and GDP measurement (Murphy 2017). In *The Economization of Life*, Murphy (2017) demonstrates how efforts to raise the GDP (an economic measurement based on numbers of people) were connected to controlling population size: economy and population were linked through demographic and contraceptive research and governmental infrastructures. The resulting demographic knowledge showed that fertility would decline as girls and young women became educated. This increasingly translated, Murphy argues, not into investments in public education, but rather into investments in girls as individuals (124).

Alongside "investment," "choice" and "rights" also figure prominently in population discourses that link reproduction to economies. Contraception and reproductive health measures are commonly represented as choices presented to households, nation states, and especially individual women, who are to be "empowered" with rights and upon whose choice and capacity to choose "correctly" rests the demographic and economic success of the nation. As MacDonald writes, "this is precisely where neoliberalism and human rights meet: at the nexus of the individual female subject as an agential citizen-consumer, full of economic potential" (2019, 267). By this population logic, as Brunson argues, choosing contraception becomes an economic imperative aimed at improving maternal health or reducing maternal mortality, and "instead of making the world safer for pregnant and birthing women, we are telling women to have fewer births" (2019, 2).

In the first sections of this chapter, I locate "investments" in population measurements, policy reports, and interventions at the global, regional, and then national scales. This means discussing two particular demographic terms – those of "demographic dividend" and "youth bulge" – in relation to priorities of reducing "adolescent fertility" and the "unmet need for contraception." In the population discourses I analyse, fertility is linked to economic development and positive futures at the global, regional, and national scales. Young women's (and not young men's) fertility is economized and even financialized; the future of the nation, region, or world comes to rest on investments made in women's choices and comportment.

In the final section of this chapter, I present young women's experiences when they became mothers in Pango Village. They experienced stress and faced scrutiny from their families as their relationships with the fathers of their children were being worked out. At the same time, much more than love between two people was at stake. The young women were blamed for "not thinking about their future." This constellation of pressures on young mothers led one to tell me that she "just wanted to be invisible" during her pregnancy. I show that the figure of the "young mother" consolidated certain anxieties about the future – namely land, marriage, and *kastom* (the Bislama term for Indigenous knowledge and practice).[2] The anxieties are in addition to the concern with access to education and wage labour.

The chapter thus also contextualizes the Children's Day parade banners on land and *kastom*, the social, political, economic, and spiritual importance of which cannot be overstated in Vanuatu. As mentioned in the introduction, land is owned by *kastom* owners, which are often kin groups. At Vanuatu's independence in 1980, all land was nominally returned to the *kastom* owners and land access subject to *kastom* law, though this was a more complex and contested process than is typically acknowledged (Rawlings 2002, 50). Legally speaking, expatriates do not own land; they rent it on decades-long leases but can demand compensation for their capital investments if the owners do not renew the lease. Land can also be leased to other Ni-Vanuatu who do not have *kastom* claims on that land (McDonnell 2016). According to Facey (quoted in Rawlings 2002, 290), in parts of southern Efate (the island where Port Vila is located), the period before independence saw a shift in land access and inheritance practices towards what people call the *blad laen* (bloodline). This term refers to inheritance between fathers and sons, and it is positioned as a replacement of inheritance through matrilineal connections associated with more expansive matrilineal social groups called *naflak*. Rawlings (2002), in his detailed analysis of land tenure systems in Pango, documents the considerable contestation over whether the *blad laen* or *naflak* should frame land ownership and access. He shows how certain families assert land claims through the *blad laen* means of inheritance, while others assert through matrilineal principles of the *naflak* (291). The question of whether *kastom* practices accept or reject women's land rights is settled in particular disputes; *kastom* is malleable and can be selectively invoked with a range of possibilities (315). In northern Efate, there has also been a significant shift to *blad laen* as the dominant way of organizing land rights and inheritance. Furthermore, McDonnell shows that, despite women having the knowledge and capacity to discuss customary land matters, certain elite men have shown themselves to be "masters of modernity" by controlling *kastom* access to land and formal registration for leases through their access to gendered spaces where land decisions are made. This has meant that the dramatic increase in long-term leases on Efate in the last decade and a half has "substantially undermined the customary land rights of women" (McDonnell 2016, 214).

What is revealed in my condensation of population measurements and discourses and local narratives and practices is the moral figure of the "young mother" or "pregnant adolescent" at particular pressure points of contemporary reproductive governance. As Lynn Morgan and Elizabeth Roberts explain, reproductive governance "refers to the mechanisms through which different historical configurations of actors – such as state, religious, and international financial institutions, NGOs, and social movements – use legislative controls, economic inducements, moral injunctions, direct coercion, and ethical incitements to produce, monitor, and control reproductive behaviors and population practices" (2012, 243). Of relevance here, the concept of reproductive governance helps us see how "some bodies become subject to monetization and financial speculation, while others may be released from regulatory oversight. The perspective of reproductive governance encourages us to consider how religious, economic, and political ideologies intersect and find expression in an array of reproductive matters" (Morgan 2019, 117). In prescribing the right and wrong kinds of conduct and investment in the reproductive capacities of young women, population discourses and local practices point to processes of reproductive governance by invoking the public moral figure of the "young mother."

Highlighting the "young mother" and related phrases at these nested scales demonstrates different ways in which reproduction is discursively and materially connected to particular futures, and the stakes involved in such framing. In population discourses, wherever children are rendered as an investment, women's reproduction is economized and individualized in association with their productive futures in wage labour. This is to be achieved through access to biomedical birth control – a public health measure – and delaying marriage. In Pango Village, a woman's education is important for her reproductive capacities for participating in wage labour. But so, too, are her reproductive capacities connected to generational obligations and maximizing her family's access to land, rendered particularly acute during a time of rapid sale of land to international occupants. Questions of marriage, because of the many exchanges, relationships, and land access renewed through the marriage process, figures prominently in the experiences and narratives of young women. In northern Efate, "land access is closely tied to marriage relationships," and the making and maintaining of these relationships can be seen as "relational, respectful, emplaced modes of women's agency" and are therefore important to women's "achievement of land rights in a climate of accelerating land leases" (McDonnell 2016, 212). Similarly, in Pango marriage is a path to respectful relations, land, and livelihood, which are difficult to achieve if someone is unmarried, making the unmarried "young mother" a particular kind of moral figure.

In what follows, I show how the "young mother" comes to matter as a social and demographic category in reference to futures in which both access to wage labour and access to land matter. The focus on monetary economies in these population discourses elides non-monetary aspects of reproductive economies, which also put pressure on young women's moral comportment. The population discourses emphasize that "the reproductive process reflects simply desire, whether for children, sex or identity fulfillment" (Bledsoe 2002, 326), and that the costs of children are limited to cash expenditures. The population discourses' focus on individuals' income potential in relation to delaying reproduction also overlooks the distributed aspects of the processes of reproduction beyond the body (Murphy 2013), something this chapter aims to highlight by considering reproduction as a distributed process that takes place in concert with marriage relations and access to land.

Global Population Publics: Girls' Low Fertility as Investment

The size, quality, and density of the world's population has been an enduring scientific concern, seen as connected with peace and planetary survival since the 1920s (Bashford 2014). Global population control became a concern in the post-war era, shaped and encouraged by international organizations and state polices promoting biomedical interventions on women's bodies to reduce fertility, with an emphasis on the

Global South (Connelly 2008). Throughout the twentieth century (e.g., Hartmann 1995; Hendrixson and Ojeda 2020) and into the twenty-first (e.g., Sasser 2014; Bhatia et al. 2020), population discourses and practices have been profoundly racialized, and indeed demographic discourses and practices have produced and reproduced racial categories (Widmer 2012, 2014). This has meant that many reproductive health interventions entangled with demographic or population data have also targeted racialized populations (Barcelos 2020).

More recently, population policies explicitly targeted at reducing births have waned. Given that those population-control policies were promoted with reference to modernity, development, and linear notions of progress (Greenhalgh 1996), one might imagine that debates over reproductive practices as a way of improving society would have largely vanished. Indeed, in population discourses, there is now an emphasis on women's empowerment and the right of the individual to control reproduction that is often articulated through neoliberal idioms (Krause and Zordo 2012). Despite this epistemic and programmatic shift, Krause and Zordo show that discipline and surveillance are nonetheless apparent in terms of "how women and couples experience stigma for not adhering to narrow norms; how accusations of irrational behavior take shape; and how new stakes for 'reasonable' and 'responsible' reproduction are cast" (2012, 142). Most recently, the emphases of population discourses connect the size and quality of the population by prescribing the need to "invest" in girls, not just women, extending from research claiming that rising rates of women's education and empowerment will translate into declining fertility rates. In such discourses, girls' education, empowerment, and reproductive health have become the focus of concern and intervention.

The main message of the United Nations Population Fund's *State of World Population 2016* report is that "the world depends on how we invest in 10-year old girls." In this report, the global institutional authority on population issues argues that investments aimed at realizing the rights and empowerment of girls will allow us to "[reap] the demographic dividend" (UNFPA 2016, 8). The "demographic dividend" is a population science term that describes a context in which the number of young people in the paid workforce grows and the number of children being born shrinks, so that people's increased earning power can be spent on fewer children (see figure 4.1). This increases wealth at the household level and economic growth at the country level "as resources that might otherwise be needed to support dependents can instead be diverted to savings and human capital" (60).

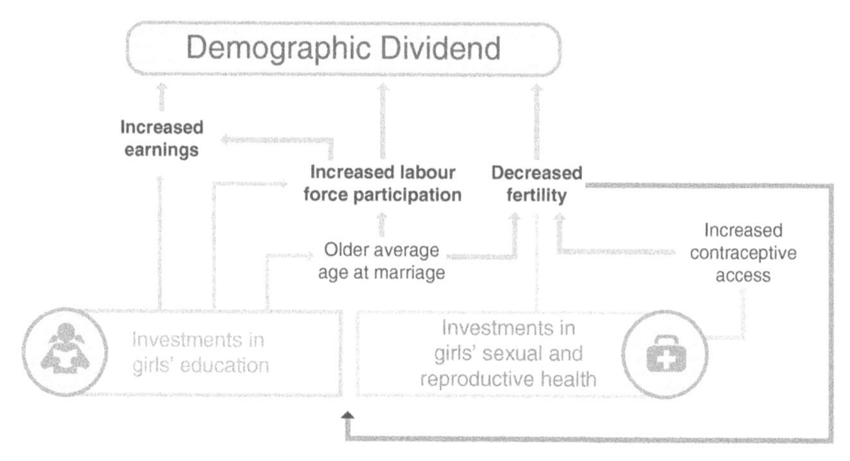

Figure 4.1. "Demographic Dividend" infographic
Source: UNFPA (2016, 50).

However, the economic gains associated with a "demographic dividend" do not automatically follow from the education and empowerment of girls. The ability to reap higher levels of income depends in part on the "human capital development" of the younger population. Young people who are healthy and educated as they reach working age have the potential to be more productive than peers who are not. Productivity also depends on access to employment and capital. A demographic dividend therefore also depends on the effective operation of labour and capital markets, institutions and policy (UNFPA 2016, 60). Realizing these economic "dividends" stemming from changing reproductive behaviours is only possible, however, when girls and young women are fully equipped to manage their fertility and empowered to make choices (8). Investments made in girls' education are twinned with investments in their sexual and reproductive health that together will lead to decreased fertility and the demographic dividend.

While figure 4.1 describes what should happen at the population level in order to achieve the demographic dividend, figure 4.2 describes the individual level by plotting the divergent paths of a girl who receives "investments" and a girl who does not.

By comparing two imagined lives, figure 4.2 shows how successful girls' empowerment and reproductive rights are expressed in the neoliberal, financialized idiom of investment, while the absence of such investment condemns another girl to a life in poverty with multiple children. The ideal presented here is a girl who completes secondary

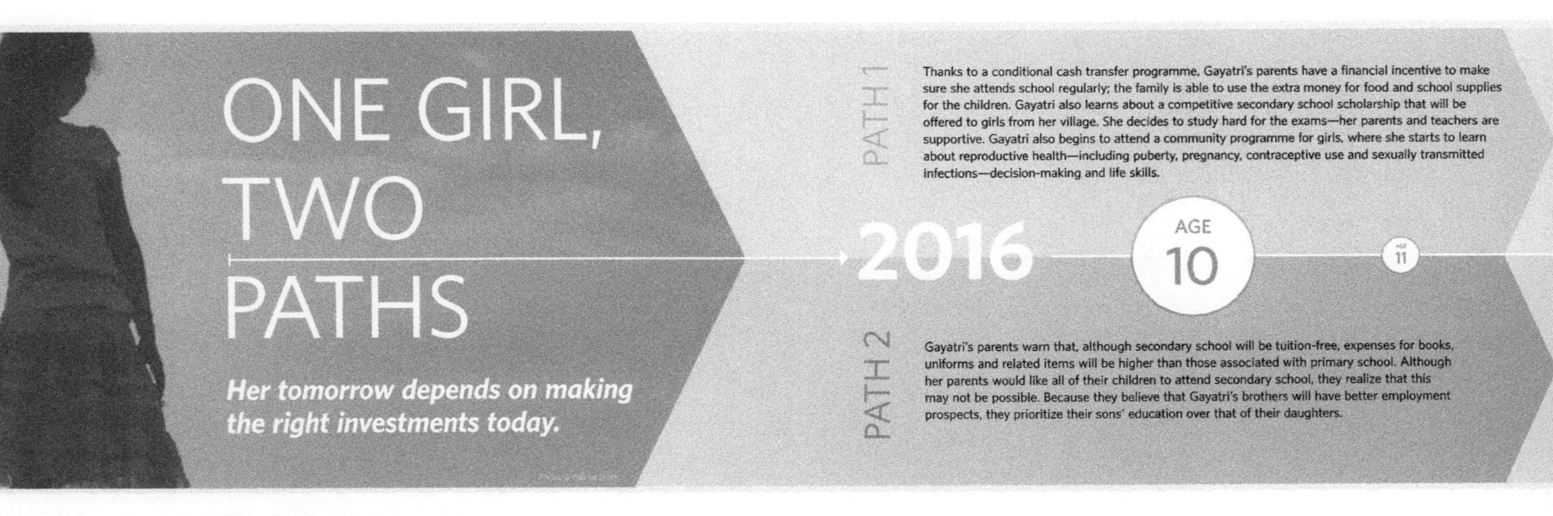

Figure 4.2. "One Girl, Two Paths" infographic
Source: UNFPA (2016, 52–5).

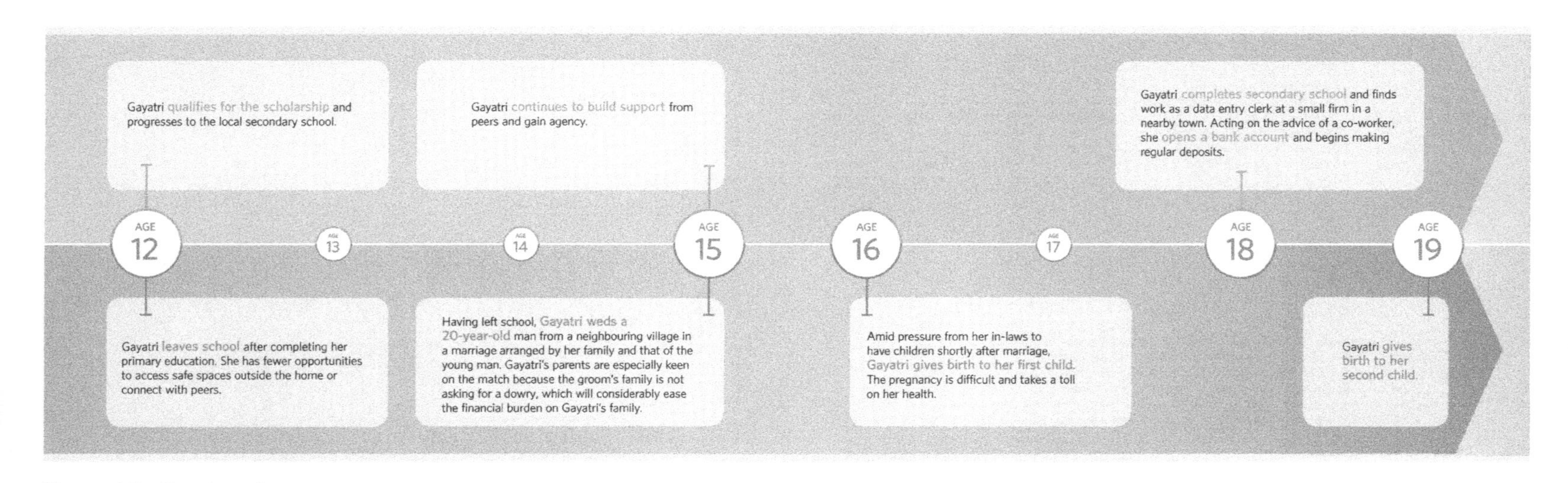

Figure 4.2. Continued

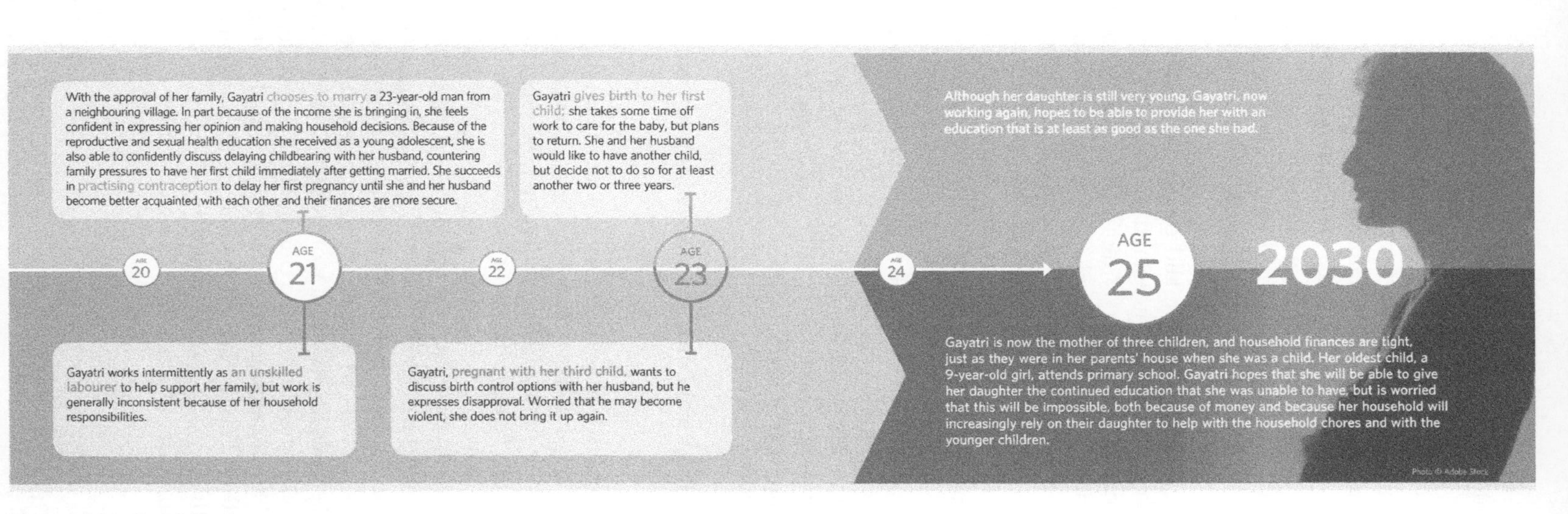

Figure 4.2. Continued

school and opens a bank account at the age of eighteen, just before her alter ego has a second child. Success is measured in individualized financial terms and smaller families. The ideal "investment scenario" pays off in higher wages, measured as a percentage of per capita GDP. Reproductive health and, even more prominently, education are indexed as solid investments, framed in moral terms for the girl and her parents as "kicking off a virtuous cycle and allowing her family to escape poverty" (UNFPA 2016, 52).

The goals to be monitored are laid out in the UNFPA's *The State of World Population 2016* report, measured by particular indicators individually tailored for each country. Among these indicators are maternal and newborn health, education, and "sexual and reproductive health." Sexual and reproductive health are to be measured by the "contraceptive prevalence rates of women 15–49," the "unmet needs for family planning rates for women 15–49," and the "proportion of demand satisfied with modern methods for women currently married 15–49" (UNFPA 2016, 106–7). The "unmet need for contraception" refers to "the proportion of women who do not want to become pregnant but are not using contraception," and it measures "the gap between women's reproductive intentions and their contraceptive behaviour" (107). It is, as I will show below, a frequently employed indicator that helps to explain the gap between knowledge, desired family size, and actual contraceptive practice.

Pacific Islands Regional Publics: Teenage Mothers and Unmet Need for Contraception

In contrast to early twentieth-century anxieties about population decline, anxieties linking social and economic problems to rapid population growth are now widespread among the island nations of the south-west Pacific. The warnings of a looming crisis escalated in the 2000s. In 2008, for example, the demographer Gerald Haberkorn wrote that the high levels of fertility, urban migration, and a lack of population policies "have the potential to derail national, regional and international development goals and objectives and, in the process, jeopardize Pacific Leaders' visions of a secure, prosperous and peaceful Pacific Region where people live free and worthwhile lives" (2007–8, 99). Similarly, in 2012, Graeme Hugo of the University of Adelaide, interviewed by the Australian Broadcasting Corporation, claimed that urban demographic growth in Papua New Guinea needs to be taken seriously, as it could eventually lead to social unrest, prompting the headline "Population Pressure Could Cause 'Melanesian Spring'" (Razak and Hugo

2012).[3] There is clearly a growing population of young people in the region, but the meaning and consequences of this fact are emergent and multifactorial.

The *State of Pacific Youth Report* (UNFPA 2019)[4] is the result of the collaboration of several prominent international organizations active in the region, and it assesses the situation for Pacific youth with regard to various indicators tied to the UN's Sustainable Development Goals. A prominent aspect of the region's demographic situation is what some have called a "youth bulge," a demographic term for a high proportion of youth in a population. It is, according to World Bank expert Justin Lin, "a common phenomenon in many developing countries, and in particular, in the least developed countries. It is often due to a stage of development where a country achieves success in reducing infant mortality but mothers still have a high fertility rate. The result is that a large share of the population is comprised of children and young adults" (2012, n.p).

In the *State of Pacific Youth Report*, the region's "youth bulge" is presented as something that can provide a "demographic dividend" if the right investments are made. The report, published by the UNFPA, also gives an account of the economic "dividend" that the region might a reap if its youth bulge is correctly managed. This exemplifies the pervasive economization of demographic processes, where patterns in the reproductive behaviour of a population are said to imply certain economic risks and opportunities. For example, with a successful demographic dividend

> families are thus able to concentrate their resources on the education, health, and well-being of fewer children. Women with a reduced child-rearing burden are more able to enter the formal labour market. Countries need to spend less on education and child health and can capitalize on increased productivity and potentially higher savings. For countries to benefit from a demographic dividend, the enabling environment must be favourable to ensure that an increased supply of workers is gainfully employed. This entails good economic management and good governance, including effective policy-making and strong institutional structures. Key institutional frameworks for education, health, the economy, and governance must be in place to convert the growing labour force into a skilled and effective workforce. (UNFPA 2019, 8)

As shown in the above passage, health services and policy responses play an important part in converting the potential hazards of the youth bulge to a demographic dividend. Realizing such a dividend depends on surplus labour being successfully absorbed into the labour market. Furthermore, the drop in fertility is the precondition of a demographic dividend. This means empowering youth to become productive workers and delay parenthood – or more explicitly, motherhood – and have fewer children than their parents' generation. An optimized youth bulge is thus to be achieved, in the language of the report, by improving "access to quality youth-friendly health services generally, and specifically for sexual and reproductive health (SRH), including addressing barriers to contraceptive services for married and unmarried sexually active youth" (UNFPA 2019, ix).

While fertility rates in the Pacific have declined, there remain "high adolescent fertility rates" (UNFPA 2019, xv). The report states that

> Teenage motherhood is particularly high in remote and rural settings. Rural adolescent fertility in the Solomon Islands and Vanuatu is 70 and 77 respectively (UNFPA 2014). On outer islands in RMI [Republic of the Marshall Islands] it reaches 100. Rates are lower in the urbanized areas of the country (80). Teenage pregnancy is indicative of higher levels of unmet need for contraception among youth, a lack of information about sexual and reproductive health and rights (SRHR), and the result of a girl's lack of autonomy over her own body (UNFPA 2013; UNFPA 2007). The pregnancy may be the result of unwanted or forced sexual relations. (UNFPA 2019, 44)

I draw attention to two key demographic measurements frequently referred to in this report, those of "adolescent fertility" and "unmet need for contraception," and how they are measured. First, the report indicates that "adolescent fertility rate or adolescent birth rate is usually defined as the number of births per 1,000 women ages 15 to 19" (UNFPA 2019, 44). Second, the "unmet need for contraception" is measured as "the percentage of women of reproductive age (18–24), either married or in a union, who have an unmet need for family planning" (xviii).[5] It is a measurement of the difference between reproductive intentions and access to, and knowledge of, contraception. Though "unmet needs" vary widely across the Pacific, "very high rates, all greater than 25 per cent, of unmet need among married girls/women 15–19-years-old are reported for Kiribati, PNG, RMI, Tonga, and Vanuatu" (56). In measurements of "adolescent fertility" and "unmet need," the focus is young female bodies and comportment, and a lack of autonomy and knowledge, making young women's health education and contraceptive use the solution to high fertility rates.

In the way that futures are projected to regional policy-making publics through demographic measurements and indicators calibrated to a regional scale, adolescent girls' empowerment is linked to access to biomedical contraception. Young women's fertility control is key to optimizing the youth bulge and realizing the demographic dividend in wage labour participation and healthy regional economic development. What little discussion there is about access to land tends to treat land as an "economic asset" that young people lack (3).

National Publics and Medicalization of Fertility Control

"When I arrived in 1949, I was asked how to have babies, when I left [in 1974] I was asked about birth control": with this summary statement, repeated more than once during our conversations in her retirement home in Auckland, New Zealand, a retired medical missionary bookended an oral history of her maternity work in the GMH on Ambae. During this same time, knowledge of biomedical contraception was changing, alongside the changes to place of birth and ideal birth caregivers that I describe in chapter 2. Looking back with the benefit of hindsight, this nun was relating her personal experience of these changes during a time of concern about population growth in Vanuatu. I would not want this statement to stand in for the ignorance of women on Ambae; rather, I mention it here because the fact that she would be asked questions of this nature at all reflects the fact that a growing number of Ni-Vanuatu women saw her medical expertise as, at the very least, another option for controlling fertility.

In the late 1960s, with population decline no longer a concern in Vanuatu, medical professionals in hospitals began prescribing medical fertility control according to availability. Two long-serving physicians at the PMH and GMH on Ambae independently decided to start offering birth control in the form of the combined oral contraceptive pill (containing both oestrogen and progesterone), and more often intrauterine devices (IUDs). In his memoir, Dr. Freeman recounts that "one of the major drives of the British administration was an attempt to limit population growth" (2006, 122). Furthermore, "It was obviously impractical to supply birth control pills to New Hebrideans ... I had just learned about the intro-uterine contraceptive device (IUCD) and, as it appeared to be a very inexpensive and effective method of birth control, I ordered some from the manufacturers in Australia" (122). Both of these methods of birth control were almost exclusively provided to married women with at least four children with the intent of reducing the stress of childbearing on the mother's body. Sometimes the husband needed to be convinced, one of the retired nurses in Pango told me.

Furthermore, the Presbyterian Church incorporated education about family planning into existing relationships and networks established over decades (still recalled by women in Pango in 2010). Dr. Freeman recalls that his wife made visits to the villages, weighed babies, and taught women about the importance of spacing children, or "family planning." These visits, Dr. Freeman writes, were "a long established tradition. They had been a means of getting to know the New Hebridean women and of providing them with both knowledge and confidence in the hospital and the mission, as well as a means of checking on the health of the babies" (2006, 124).

In the 1970s, Depo-Provera injections were available in the areas around Vila, and, as one prominent expert on family planning in Vanuatu told me in an oral history interview, these were reportedly at times given to women without their full consent or knowledge. After independence, when the Vanuatu government took over the PMH, medical professionals stopped distributing Depo-Provera in this way, because an influential Ni-Vanuatu politician, appalled by the lack of information that accompanied these interventions, took control of the situation. Abortion was officially banned in Vanuatu's legislation after independence. In 1990, the Vanuatu Family Health Association began its work. Its members trained peer educators, did radio interviews, and put up posters, all in the name of providing more information. In the Vila area, the churches facilitated the association's work and the chiefs did not oppose it. At times, the Vanuatu Family Health Association faced opposition because of rumours that it was stopping people from having children. In reality, according to a former staff member, members were only talking about family planning, and only after three children. From the beginning, biomedical contraception was in short supply, trained staff were limited, and such contraception was located in the context of marriage and families and made available to married women with the goal of providing them with choices that would support their health.

Around the time that doctors began distributing biomedical contraception, researchers (e.g., Bedford 1973; Bonnemaison 1976) began worrying that the population of Vanuatu, especially its urban segment, was growing too rapidly. Bedford and Bonnemaison voiced concerns that urban development needed to be sensitive to Indigenous land use practices and cosmologies. Bonnemaison in particular was concerned about the eventual creation of urban poverty and informal settlements. By the time I was conducting research in Vanuatu in the early twenty-first century, concerns about population growth had grown more acute. VCH staffer Robert F. Grace declared, "If this population growth continues, it will place intolerable strains on the development of the country"

(2002, 17). Most alarming were Helen Ware's (2005) predictions that, due to population growth, Vanuatu stood to follow in the path of political violence down which the Solomon Islands and Fiji had travelled. In 2012, the UNFPA claimed that Vanuatu's "total fertility rate of 3.9 lifetime births per woman is one of the highest in the Pacific" (2012, n.p.). The UNFPA also explained that, "with a current population of approximately 271 thousand, Vanuatu shares with neighbouring Solomon Islands the distinction of having one of the highest rates of population growth in the Pacific (2.4 and 2.3 percent annual growth, respectively)" (2014, 95). All demographers worried about the large number of young people who were, or soon would be, having children and who did not have jobs, the unequal position of women, urban migration, and the lack of access to birth control. Proposed solutions included increasing tubal ligations and barrier methods of contraception (Grace 2002), increasing the education of women (Grace 2002; Haberkorn 2007–8; Ware 2005), and promoting emigration as a "safety valve" (Ware 2005, 451). The Pacific Institute of Public Policy, based in Port Vila, described the demographic situation as a potential threat to democracy. In this view, population growth was less immediately a problem of fertility than one of land distribution:

> Young people are getting angrier at the increasing alienation of their birth right – their land. There are more and more young people, increasingly urbanized, many without *kastom* land to return to and few opportunities for formal employment. Higher costs of living, rising food prices and land being sold off to foreigners are all adding to the pressure. What happens when a couple of generations have no jobs and no land? (Pacific Institute of Public Policy 2011, 4)

Around this same time, in June 2011, then prime minister Sato Kilman launched, with the financial support of the UNFPA and the Secretariat of the Pacific Community (now Pacific Community), the National Population Policy. The cover of that document shows the statue of the family group that stands outside the national parliament. It is decidedly a nuclear family: father, mother, and two children. In his foreword, Prime Minister Kilman states clearly that "the National Population Policy 2011–2020 is not a 'population control' policy, it does not aim to tell families how many children they can have nor does it attempt to control population movements within the country" (Vanuatu Department of Strategic Policy, Planning and Aid Coordination 2011, ii). Rather, he argues, understanding population is key for economic prosperity. Kilman continues:

> Knowledge of population change is essential for effective development planning whether at the national or provincial level. It is clear that for the next several decades our population will continue to grow, and this growth must be taken into account in all our planning activities. Effective development planning will, in turn, lead to faster economic growth and improvements in our social indicators. In the longer run this will reduce the rate of population growth to a more manageable level. (ii)

The first goal of the National Population Policy is to "reduce fertility and unintended pregnancy particularly among target population groups" (Vanuatu Department of Strategic Policy, Planning and Aid Coordination 2011, 44);[6] the main target is "teenage pregnancy," particularly in rural areas. The narrative description (in the *Demographic and Health Survey 2013*, discussed at length below) concerning adolescent fertility in the population policy further explains why this focus is required:

> Childbearing by adolescents has potentially negative demographic and social consequences. One of the key components of Vanuatu's population policy is to reduce overall fertility and focus on teenage pregnancy. Children born to very young mothers tend to be predisposed to a higher risk of illness and death. Also, teenage mothers are more likely to experience complications during pregnancy and are less likely to be prepared to deal with such complications, which often lead to morbidities or even maternal death. From a social perspective, it is to be noted that early entry into reproduction denies young women the opportunity to pursue academic or working careers. Consequently, younger mothers tend to have less education and lower earning potential. Finally, the psychological immaturity that characterizes most teenagers is likely to have detrimental effects on the wellbeing of both mother and child. (Vanuatu Ministry of Health, VNSO, and SPC 2014, 58)

Like the demographic measurement of the "unmet need for contraception," which focuses on female actions, the "teenage mother" or "pregnant adolescent" also concentrates on the female body in order to prevent the "potentially negative demographic or social consequences." The following strategies are proposed to meet the goal of reducing fertility:

- Implementation of RH [reproductive health] strategy at all levels (national, provincial and area council) applying the primary health care approach with adequate resourcing;

- Improve access to RH services including family planning applying gender responsive and human rights approaches;
- Improved access to RH services and knowledge for specific target groups such as men, adolescents, single mothers and other vulnerable groups;
- Integration of Health and Family life education in the school curriculum;
- Community education and health promotion on family planning to increase contraceptive use;
- Strengthen multi-sectoral partnerships of line ministries and other relevant stakeholders. (Vanuatu Department of Strategic Policy, Planning and Aid Coordination 2011, 45)

Throughout the policy there is a concern with matching the growth of the population to appropriate development. In this respect, the policy focuses on providing education for all, but especially girls, in accordance with the second Millennium Development Goal[7] (that of achieving universal primary education), and as a necessary precondition of economic growth. Social development is recognized as important for economic development, and it is under this heading in particular that women's access to land is addressed: "There is also the issue of 'kastom', which tends to limit the role of women particularly in the context of decision-making at the village level and the ownership and control of land" (Vanuatu Department of Strategic Policy, Planning and Aid Coordination 2011, 11). Land use planning does appear in the document, though not in terms of access and women's empowerment, but rather as a strategy for managing rural–urban migration and urbanization.

In a ground-breaking report based on research conducted in 2013 and entitled *Demographic and Health Survey 2013* (Vanuatu Ministry of Health, VNSO, and SPC 2014), Vanuatu government researchers asked men and women about their knowledge of and experiences with family planning and their views on ideal family size; they also asked about health issues relating to pregnancy, childbirth, childhood nutrition, and infectious diseases. The survey is a remarkable achievement in terms of its ability to make reproductive knowledge public in Vanuatu. In 2010, I heard from nurses, both working and retired, that knowledge of contraception and sexuality is "*tabu*" and "secret," making these subjects very hard to talk about. During interviews, these nurses would often gesture towards me and say, "white people like you, you talk about this, in Vanuatu it isn't talked about." *Demographic and Health Survey 2013* suggests a shift in public expectations when it comes to speaking about pregnancy and family planning.

The survey aimed to provide a picture of Ni-Vanuatu knowledge and use of contraception, both modern and traditional (terms used in the report). Women who said they had heard of a method of family planning were asked whether they had ever used that method. Researchers asked men whether "they had ever used 'male-oriented' methods, such as male sterilization, condoms, rhythm or withdrawal" (Vanuatu Ministry of Health, VNSO, and SPC 2014, 62). As for the rates of use, the researchers wrote that contraceptive use is higher for married women: 49% reported using any method, and 37% reported using a modern method (66). Further,

> The overall contraceptive prevalence rate among ... all women is 38%, with 29% using a modern method and 9% using a traditional method. The most widely used methods of contraception include female sterilization (8%), birth control pills (8%) and injectable contraceptives (7%), followed by the rhythm method, which is used by 5% of all women. Modern contraceptive use for all women rises with age, [and] peaks at 40% among women aged 30–34, and then fluctuates. (66)

The survey found that both men and women's knowledge of contraception was high, with 91 per cent of all women and 98 per cent of all men aged fifteen to forty-nine knowing of at least one method. Biomedical contraceptive methods were most widely known: 90 per cent of all women knew of at least one method, while 62 per cent knew of a traditional method. Desired family sizes were also researched, and usually women wanted smaller families than men.

Reflecting widely accepted research practices in demographic analysis, seen at previous scales, the survey presents fertility as connected to educational and financial status, with both variables influencing family planning and contraceptive practices. The survey's design linked fertility to knowledge and access to contraception as well as women's empowerment. Researchers assessed women's empowerment through "direct measures of women's autonomy and status" (Vanuatu Ministry of Health, VNSO, and SPC 2014, 225). To measure these, "questions were asked about women's participation in household decision-making, their acceptance of wife beating and their opinions about the circumstances under which a woman is justified in refusing to have sexual intercourse with her husband or partner" (225). Women's autonomy was also measured by asking women about their control over their own incomes and the degree of control they have over household finances (225).

Notably, researchers enquired about men's, and not only women's, attitudes towards marriage and family size and their knowledge of contraception; but in the demographic indicator of "unmet need for contraception," it is women's rates of contraceptive need that were tracked. This is significant because such indicators ultimately carry more weight and therefore result in more substantive directives for practical action.

My intent here is to consider what, exactly, the population policy and survey make visible by making reproduction public. At the scale of national-level population discourse, policy and research lay out interventions that emphasize young women and the explicit economization of their fertility and education. There is very limited mention in the survey of either men's or women's access to land and the fulfilment of (non-monetized) collective potential. It draws a connection between controlling reproduction and individual empowerment, education, and access to cash wages. In turn, individual empowerment is said to lead to positive national economic development. In its suggested priorities and points of intervention, the 2011 policy renders the economic prosperity of the nation as an effect of the reproductive choices and education of young women, who play a limited role in village decision making and have limited access to land.

For global, regional, and national publics, the population discourses around family planning generate the closely related figures of the "young mother" and the "pregnant adolescent." These figures are an assemblage of demographic indicators, policy objectives, and technological solutions. At all three scales, women's reproductive capacities are associated with strong national economies. The research categories, metrics, indicators, and policies of the population reports mean that young mothers appear as the focus of reproductive governance; young women's educational empowerment, fertility control, and cash-earning potential are yoked to the success of global, regional, and national futures. Their access to land is mentioned in one place as a challenge that *kastom* often fails to meet.

At three nested scales, population science's focus on women's access to education and the available technologies of contraception frame the teenage female as the point of intervention. There are no demographic indicators that focus on problematic "teenage fathers" or young men with "unmet contraception needs." This is not to demonize the undoubtedly good intentions of any population scientists, but rather to show that an unintended consequence of these measurements is that the reproductive behaviour of a society, and that society's capacity to manage that reproductive behaviour, is reduced to a matter of women's bodies and reproductive choices. It is also to point out that reproduction and human empowerment are linked in ways that prioritize monetary value over many other concerns, as demographic and population discourses render goods like health and education as investments.

Young Mothers as Social Identity and Village Experience

The Children's Day parade that I described at the opening of this chapter was a remarkable public display of positive feelings about children, but at other times the large number of children being born was also acknowledged as cause for concern. In terms of demographic statistics, the total fertility rate in 2013 was 3.4 births per woman. The population's 2020 annual population growth rate was 2.4 per cent, down slightly from 2.6 per cent in 2010. Life expectancy was 52.4 years in 1970, 71.3 years in 2012 (UNICEF 2013), and 74.47 years in 2019 (World Bank 2020).

One way that alarm over the increasing number of births circulated in Pango was in rumours concerning the high number of births per month at the VCH. Stories of new mothers having to sleep on mats on the hospital floor circulated widely. These accounts, which register real fears about a lack of hospital resources (a structural problem with colonial origins, as described in chapter 2), frequently invoked the sheer number of "young mothers" as a central factor underlying an overtaxed health-care system. The figure of the "young mother" in these stories was a salient social identity evoking both pity and harsh judgment. For example, young mothers were on the one hand the subject of fervent prayers on Mother's Day in the Presbyterian Church; on the other, they were cast as selfish and reckless and mentioned along with other social problems associated with youth, like excessive kava or drug use.

When I enquired about the factors that made young mothers so challenging, I heard – especially from older generations – that the fact that these mothers were young was a problem, but equally difficult, if not more so, was the fact that they were unmarried. There is a perception that this is a new phenomenon associated with modern life, and symptomatic of a broader moral decay and the breakdown of obligations between generations and traditional authority. Young mothers were reproached for being reckless, for not thinking about the future or "the consequences of their actions." They were described as stubborn (*strong hed*) and selfish.[8] "Selfish," in Pango, means many things, but in this particular context it refers to a neglect of social obligations and relationships that link the current generation of young adults with older ones (see also Lind 2014).

The social identity of "youth," Mitchell writes, has shifting meanings, as on the one hand they are considered the "most vulnerable victims" who cannot find paid employment, while on the other they are the "perpetrators of disorder" who no longer respect traditional authority (2011b, 38). Relatedly, Cummings has found that there is a "gendered culpability" in how "women more often than men are deemed culpable for the negative effects of social change" (2008, 134). The social identity of the "young mother" is at the intersections of both gendered culpability and representations of youth as troublemakers.

What was it like to become a "young mother"? To learn about the pre- and postnatal experiences of young women in Pango, I walked on dirt paths through the village over the course of many afternoons in 2010, accompanied by my friend and research assistant. We passed children playing outside houses made of corrugated iron, houses made from traditional bush materials, and new cinder block houses with indoor showers and toilets. Together we spoke with sixteen mothers whose children were under one year of age, twelve of whom were not married to their child's father and lived with their parents and extended families.[9] Afternoon being a time when there was a lull in the never-ending duties of laundering, cooking, and gardening, we sat on pandanus mats under shelters outside corrugated iron houses, at tables, or on couches in cement houses and listened to the stories of our interlocutors.

Virtually all of these unmarried young women were supported, financially and otherwise, by their parents, and most lived on land owned by their father's kin group, or in houses rented with family members if they were from another island. Starting their own household would have been unthinkable financially and socially for these women, so the question of marriage entailed a great deal for them, their future children, and their families of origin. The married mothers we spoke with typically lived in their own house with their husband on land owned by their husband's family. With a few exceptions, it was only married men that financially supported their children. The mothers we spoke to who were unmarried at the time of their first child faced significant uncertainty about how the immediate future would unfold. Our open questions about their experiences during pregnancy would often turn to more focused discussion on the future involvement of the young man and his family, and particularly on whether a marriage would take place. While the relationships between the unmarried young mother, the biological father, and the baby were being worked out, the young woman experienced what was often described as a painful form of social isolation. The ensuing discussions she would have with the young man would shape far more than the love she felt for him; it had ramifications for her child and extended family as well. These discussions would come to bear on the land she will have access to for the next phase of her life, and therefore the future of her children.

When discussing their experiences of pregnancy, birth, and early post-partum time, many of the women, but especially the unmarried mothers, talked about how they would *stap kwaet* (be quiet) and not go out much during their pregnancy. The young women feared the anger of their own parents (who generally welcomed the baby once it was born). Learning when to *stap kwaet*, writes Cummings (2008), is an aspect of managing feminine sexual modesty in Vanuatu. For young pregnant women who were unmarried – and through oral histories, I found this is not a new phenomenon – avoiding public scrutiny was a prime concern during pregnancy.

There were many reasons why women might *stap kwaet* and be careful not to *wokbaot* during pregnancy. One reason is that pregnant women are thought to be vulnerable to sorcery (see also Bourdy and Walter 1992, 187), and so are encouraged to stay close to home and avoid places where sorcery attacks could happen. Sorcery is suspected in situations that appear unusual (and might thus lead to illness), and for most illnesses generally. Seeing a snake in an unusual place, for example, would give rise to the possibility of being influenced by sorcery in the near future, and the individual concerned would then pay attention to relationships that might not be reciprocal. Many people recognize sorcery's potential for influencing particular aspects of life in Vanuatu. For example, it has been said to intentionally make someone fall in love, make someone forget important information, make someone ill, cause someone harm out of jealousy, or cause bad weather in order to ruin an event (see also Mitchell 2011b; Rio 2002).

Pregnant women are aware of the sanction to avoid, at any time of day, those dangerous places in the landscape (*tabu ples*) where their clan's (*naflak*) ancestral spirits are present, lest the fetus be harmed. One cautionary tale circulated in a viral text message that was a frequent topic of conversation; it related the experience of a woman on another island who had given birth to a deformed baby that looked like a snake. The assumption was that she had violated some *tabu*. In the face of such possibilities, pregnant women who *wokbaot* (and certainly people do) run the risk of being seen as culpable if their pregnancy or their child's health is compromised in any way.

This feeling that they needed to *stap kwaet*, as reported by interviewees, did not appear to negatively impact women's ability or willingness to access biomedical health care, though in a few cases it delayed their choice to do so. The young women all made use of the prenatal care available at the VCH, had ultrasounds, and took anti-malarial and iron tablets. Fear of gossip did affect their access to family planning, in that the fear of being seen at the clinic kept young women from going to get information and supplies, a point reiterated in my many conversations with community workers, peer health educators, and family planning counsellors (see also Cummings 2008). More than half of the young women interviewed said they first received information about family planning at their initial post-partum visit; most said they wished they had received more of this information sooner, identifying this as a problem that needed to be solved. When family planning information is distributed in Vanuatu, Servy (2018, 43) demonstrates that it comes with moralistic narratives that organize the possible methods hierarchically, from practising abstinence, to maintaining fidelity in one's relationship, to using a condom.

In contrast to narratives claiming that young women are irresponsible and do not care about what others think, the interview responses suggest that they very much feared being chastised for having apparently not thought about "the consequences" of their actions or having "ruined their future" (*spoilem fiuja*). Many would have appreciated easier access to birth control and greater support from the fathers of their children. The young women also worried about the reactions of the parents of prospective fathers who, according to the selectively invoked *kastom*, were not obligated to acknowledge any children unless the parents had entered into marriage.

The young women reported that these parents would not chastise (*tok strong*) their sons in the same way that they would a young and unmarried pregnant daughter. It is also worth noting that there really is no social category of "young father" garnering the same moral concern. When we talked about this discrepancy between "young mothers" and "young fathers" with the young women – the latter term eliciting smiles and giggles from the interviewees – we typically heard that *kastom* puts men at a higher position, that women are *strong hed* and do not think about the consequences of their actions, and that, because of *kastom*, the parents of an (unmarried) girl and young mother have to look after her and her baby.

There is a broadly held hope, among those families who can financially manage to keep their children in school past year six, that young women will finish school before having children because children will prevent the woman from doing so. Still, after becoming mothers, the young women I knew did not stop working to make a good future for themselves and their children after this life transition. Young mothers participated in subsistence agriculture and paid and unpaid household labour and childcare, making it possible for other parents in their household to engage in paid work. It was possible to earn cash by one's own initiative, and many did this by cooking food like *tuluk* (a popular snack of baked manioc dough wrapped around shredded pork), popcorn, or cake and selling it around the village, or by making small necklaces for tourists. Young women could also work as a *haos gel* (domestic worker) for families in Pango or Vila. Higher-paid and higher-status jobs like working in a bank, in the civil service, or for an NGO were also jobs undertaken by women in Pango, though these women were in the minority among our interviewees. These productive activities were more feasible for those who had a greater variety of supports, mostly from their parents and siblings. In turn, their paid work contributed to the household.

"The Only *Kastom* Left Is Marriage"

"The only *kastom* left is marriage," an older man told me in a conversation about the history of the village. This statement was frequently repeated by members of the oldest generation. As evidence of *kastom*, older people in Pango told me that *braed praes* (bride price) is generally paid (depending on the Christian denomination) and the groom's family gives cash and traditional wealth items like yams, pigs, and mats to the bride's family. In historical narratives in Pango, women are described as peacemakers between villages, families, and *naflak* through marriage arrangements. In ideal situations, a woman would go to her father's *naflak*, and a woman from that *naflak* would then marry into hers. While these unions were possible to break and were not universally practised in previous generations, they were much more common than they are today.

In contemporary marriages, the union is agreed upon by the parents of both the groom and the bride and the engagement is then formalized in a ceremony. In the months leading to the wedding, there are meetings of family members at which it is decided how the wedding will be paid for, and this creates relationships between the kin groups. In 2010, the preparations for a *braed praes* ceremony, a church wedding, and a post-wedding feast involved many, if not most, families in the village. The women were occupied with sewing dresses, making mats and baskets, and preparing food, while the men were engaged in making shelter and readying the pig to be cooked. There were frequent trips into town to buy what could not be had in the gardens. The objects and amounts of money presented at the wedding are meticulously listed in a book, which families can then consult in order to provide proper reciprocity at future weddings. Such transactions are often presented by international development organizations as curtailing women's rights and reproductive autonomy, but as Servy shows of *braed praes* in Seaside Tongoa (an area in Port Vila), the payments are "mainly used to create protective and supportive relationships between social groups, rather than to give rights over women's productive and reproductive capacities" (2020, 307). The *braed praes* also makes relations that bear on land access and inheritance public in North Efate (McDonnell 2016, 212).

"You will be the last [anthropologist] to go around and hear about the *naflak*!" a senior woman told me. People would repeat this claim: the *naflak* is not very important anymore, and now the *blad laen* is important. So, while the traditional story still circulates about women being *rod* (roads) between *naflak* through marriage, a respectable path in 2010, if not to marry into another *naflak*, was for a woman to marry a Pango man and go and live with his family on their land. If she has a child before the marriage, different solutions are brought to bear in order to clarify the links the child has to previous generations. Sometimes, members of the groom's family would *pemaot pikinini* (give objects in exchange for a child) when they gave the *braed praes*. In such cases, the groom's family would identify a portion of the gifts as specifically identified with the child of the bride, thereby clarifying the inheritance rights of the child.

"Culture Has Us by the Throat": Marriage and Land

I also listened to senior women's oral histories of what it was like to raise their children in Pango in the 1960s and 1970s. Marriage and land use figured as structuring idioms in oral histories of the village and of individuals' lives. I listened as one of my interlocutors told the village history according to patterns of collective land use: "We [Pango people] used to live closer together, down by the well, but now we live more spread out." An individual's life stages were similarly marked by phrases like, "We [the family she was born into] lived down by the well when I was a child, and then I moved here [to husband's land, or land the woman's father gave her] when I got married." I deliberately did not ask direct questions about who owns what land or about specific land tenure practices: this would have shut down these conversations immediately as there had recently been some large disputes over land sales and chiefly title; nor was this part of the research permit I had from the Vanuatu Cultural Centre or my research agreement with the chief of Pango. I quickly learned that even asking questions about genealogies could curtail conversations. Though I did not directly ask about land, access to land arose frequently in the birth stories of older women, and in conversations about marriage for younger women.

As mentioned earlier, many people told me that marriage traditionally meant that women were *rods* between two families or clans, but from the vantage point of young women in 2010, marriage was a road to status, love, respectability, and clear land inheritance for one's children. For young women who became mothers, the way to acquire a place to live – both in the sense of land to grow food as well as a house to live in – is either by marriage or by the good graces of one's own family.

"Culture has us by the throat!" a middle-aged woman with adult children told me passionately as she neared the end of her oral life history. She was a talented woman with advanced professional education. Her father had arranged a conventional marriage for her, as a way of making peace with another *naflak*, and she had subsequently endured intimate partner violence (as did many women of her age), got divorced, and was then at the mercy of the men of her generation for access to land. She was still struggling to gain access to the land of her *naflak*, what she called "our mother's land," which others said was now under control of another *blad laen*. This required negotiating and maintaining a web of relations and following the expectations people had of her, a challenge she struggled to meet, despite her best efforts.

The frustration this woman had experienced over the course of some decades taught me a lot about the contemporary concern of the "young mother" about whether her baby's father would acknowledge the baby and whether he and his family would agree to a marriage. Women's and mothers' conduct has long been viewed in Vanuatu as a synecdoche for these women's families, their villages, or the nation at large. Women's activities were the focus of Presbyterian missionary endeavours in the late nineteenth century (Jolly 1991), and women's mothering abilities were a key interest of researchers during the time of population decline (Jolly 1998, 2001, 2002). In a particular articulation of tradition in northern Vanuatu, the range of acceptable comportment for women was curtailed by the nature of men's labour migration in the twentieth century (Jolly 1987).

In this context marked by an increase in *blad laen* shaping land access, an intensification of land use, and an increase in the long-term leasing of land to resorts and expatriates, the comportment of the unmarried "young mother" has become the object of intense scrutiny and debate, the stakes of which – for both the mother and her family – feel higher than ever. The public identity of the "young mother" therefore emerges as a moral figure in stories of reproduction in a way that underlines reproduction's dependence not only on biological processes, but also on continuously negotiated relationships of kinship, property, and social obligation.

The public narratives in Pango about "young mothers" who do not think about the future need to be understood in the context of larger anxieties about the growing pressure that capitalist social relations place on land use and kinship relations, which serve to demonstrate the current entanglement of reproduction and monetary economies in two ways. First, the concern that young mothers might be prevented from completing their education is related to concerns about their ability to participate in wage labour, an increasing necessity. Second, with the number of resorts in the vicinity of the village increasing, and the amount of land being transferred to expatriate leaseholders growing, a more tightly constrained supply of local land has underlined the importance of marriage as a means for young women to access land. The figure of the "young mother" therefore demonstrates importance of marriage in relation to land in Pango. These pressures have led some women to want to be invisible, and others, who are brave enough to say it, to say in different ways that "culture has us by the throat." It also means that "young mothers," who are generally unmarried, are implicitly a threat to collective futures, unlike married mothers, who have clarified their and their children's access to land and their husbands' kin through marriage.

I raise women's frustrations with culture and the more general anxiety around "young mothers" in Pango as examples of what a broad focus on reproduction can show us about the entanglement of wage-based and traditional economies that exceed the economization of the fertility of the "young mother" in population discourses. The opinions and experiences shared by these women show how a monetary economy and a social and legal context in which people achieve access to land through kinship ties come together to create distinct power relations during times of population growth and dramatic shifts in land use practices. On the one hand, there is increased pressure to earn wages or cash because many necessities must be purchased in town or paid for with cash (like electricity and water). So young women are encouraged to stay in school in order to position themselves for the best possible jobs. On the other hand, access to land is of increasing importance as the population grows and more land is leased. Women have more options in this respect through marriage. These concerns with land, marriage, and *kastom* demonstrate the complex relations of reproduction in Pango that cannot be accounted for by framing reproduction only in terms of women's economic empowerment, access to wage labour, and biomedical contraception – the factors that demographic and population discourses tend to emphasize.

Anna Naupa (2017), a researcher and gender and land policy expert for the Vanuatu government, writes that, because men can assert power more readily, it has been a challenging process to advocate for women's continued *kastom* access to land in the context of an increasing number of land leases. She argues that "future advocacy efforts [for women's land access] must include greater engagement by women themselves, not just their advocates, for reform efforts to be sustainable" (306).

Conclusion

In this chapter, I have shown how the "young mother" emerges as a key moral figure in how reproduction is made public. In population discourses, young women's empowerment and access to contraception are treated as universally important and yoked to the future economic prospects of the globe, the region, the nation, and the individual. These women's reproductive empowerment is connected to investments made in their capacity to earn cash. Lives are valuable when money is made, not when they are dependent on others. To put this in terms of reproductive governance, the promise of escaping poverty, both individually and as a nation, region, or globe, is presented as a means of controlling or governing young women's reproductive behaviour, an important investment. This is the moral figure that focuses public attention on reproduction in individual bodies and is the focus of public policy, development agendas, and public health interventions.

At the village level, the "young mother" is a moral figure whose problems are to be overcome by marriage. In this respect, young mothers, particularly those who are not married, are a focus because of the social and legal processes by which land access is achieved through marriage or kin relations. In the past, access to land via *naflak* relations meant that women (and men) must access land via links to their mothers, but now access through *naflak* or *blad laen* can be a source of dispute. Marriage means that women and their children's place in a web of relations is publicly acknowledged, thereby giving them more options for achieving access to land.

The concern about the selfish comportment of "young mothers," at the village level, reveals anxieties about a future in which wage labour is seen as more and more important and land access more dear. In this future, women need access to education and also access to land. Given that access to land for most women in Vanuatu is acquired not through purchase or lease, but through kinship ties, the pressure to clarify one's access to land through marriage is intense. At the same time, population discourses and policies make young women and girls more and more visible as targets who can improve themselves and their communities' socio-economic situation more broadly, if they pursue education, delay marriage in favour of wage labour, and have access to biomedical contraception. While education, cash employment, and biomedical contraception are crucially important, they cannot compensate for the reliable base of material subsistence that access to land represents in Vanuatu. The women who become mothers before marriage feel this in acute ways, and therefore often choose to *stap kwaet lo haos nomo* (just stay home) or say that they "just wanted to be invisible."

Chapter Five

"Well-Being for Melanesia": Alternative Indicators, Massage Healers, and Knowledge of Reciprocal Relationships, 2010–2021

The future of Vanuatu will very much depend on what approach the government decides to take. If Vanuatu decides to imitate other countries of the world, there can be no freedom in terms of being one's own master with one's own individual identity. But in deciding to be truly independent from any other country, whether within the region or afar, we shall have to work even harder to achieve this. The main effort will then be to really polish up our very own Pacific and Melanesian ideas, to make them the basis of unity in our own country and within our region and to give us the necessary strength and direction to choose wisely what we want and do not want for the future.

Father Walter Lini, first prime minister of Vanuatu, at independence in 1980 (text from the concluding minute of the *Well-Being in Vanuatu* video; MNCC and VNSO 2012)

Joseph set his fishing lines, which stretched to the edge of the reef, while my family and I sat nearby, facing the open ocean.[1] When he had finished, he gestured to his right at the as-yet undeveloped, pandanus-lined shore that stretches towards Pango point, and described how much land had been acquired by expatriates in recent years. He told us how little money people receive in annual rent from the resort owners, who hold seventy-five-year leases on the land.[2] At the end of a lease, should the traditional land owners want their land back, they would have to compensate the renter for any capital that they had invested in the property. More often than not, this meant that the land was effectively sold. "In the future," Joseph said, "Pango people will be begging in the street." His suggestion that people in Vanuatu would be reduced to public begging – a practice virtually unknown in either urban or rural areas of Vanuatu – marks the scale and depth of the change that Joseph was describing. The prospect of public begging in Vanuatu expresses an anxious awareness of the possibility of lives being lived in direct opposition to local notions of well-being and of the complete erosion of the basis of social reproduction.

Vanuatu is a place where lives are lived through local exchanges and obligations that depend upon a complex of social and environmental networks that include and exceed capitalist frameworks. It is a place where, "out of a population of [approximately] 270,000, 220,000, minimum, aren't in the formal economy" (St-Hilaire of the Vanuatu Financial Centre Association, quoted in Bremner 2017, n.p). As I showed in chapter 3, a similar statistical picture of local economic activity impressed the first census takers in 1967, which they enumerated through the figure of "subsistence." The obligations existing outside of the formal economy have shifted and yet persisted through forms of colonialism, capitalism, and Christianity that Ni-Vanuatu adopted. Joseph's fear that people might resort to begging on the street is an expression of what it means to live in a peri-urban capitalist frontier in the twenty-first century, a place where the population is growing, where land use and labour patterns are changing, and where people worry about future forms of care. Begging is the result of poverty, a state and relationship that is almost incompatible with the core value of reciprocity that is central to social reproduction in Vanuatu.

In a context in which fundamental Ni-Vanuatu relationships of social reproduction exist under such existential threat, what forms of knowledge and care are mobilized? In this chapter, I address this question by looking at two ways of expressing knowledge of well-being. First, I discuss the "alternative indicators of well-being for Melanesia," which were developed, researched, and published by Ni-Vanuatu politicians, chiefs, and statisticians, as well as an overseas researcher. Second, I analyse the knowledge held by women massage healers, who tend to the needs of most, if not all, pregnant women in the peri-urban context of Pango Village.[3] The first case presents a strategic and quantified public engagement effort, adopting the logic of indicators and modernity, and producing a moral figure of "well-being for Melanesia." The second case relies on an intensely local form of expertise and understanding of well-being, a set of normative ideals that allow people to identify features of the landscape and reciprocal relationships between humans and between humans and non-humans, which in turn presents its own moral figure in the massage healer. I argue that each of these two modes and expressions of Ni-Vanuatu knowledge of well-being demonstrate how well-being is tied to an ideal of reciprocity. One form of knowledge employs quantification to make social reproduction public, while the other, which is closely connected with care for reproduction, is not rendered public, even in the alternative indicators of well-being. They are both assertions of Indigenous futurities, of how to lead a good life that is Melanesian, or Ni-Vanuatu, in the face of struggles presented by capitalism and international development.

By putting both kinds of knowledge in dialogue in this chapter, I hope to do several things. The first is to show how the indicators of "Melanesian well-being" assert quantified knowledge of social reproduction for well-being as a form of sovereignty. Second, I want to explore what the public quantifications of social reproduction downplay as a consequence of their economistic orientation, and particularly their inattention to women's care practices and knowledge of care for well-being. Examined together, these two scales of knowledge show the importance of reciprocity in Ni-Vanuatu understandings of well-being, whether in the context of proactive engagement with capitalist social forms at a national scale for social policy, or at a very localized scale in the practices of care and well-being that exceed these capitalist forms. When audit culture and metrics are combined with Ni-Vanuatu ingenuity, the audit culture and metric social forms can be spaces for Indigenous futurities, which are also articulated in the massage healers' practices of care that exceed economistic accounting. In this sense, I want to touch on the limits of audit culture and metrics, and in so doing highlight Ni-Vanuatu capacity to imagine sovereign futures.

The association of knowledge with power through effective control and awareness of an audience is a long-standing aspect of Ni-Vanuatu societies (Lindstrom 1990). Lindstrom argues that Ni-Vanuatu societies are knowledge-based in ways that trouble the naturalized association of knowledge-based societies and post-industrial economies. Beyond their own societies, Ni-Vanuatu have been conscripted into conversations over the nature of knowledge for decades. In the late nineteenth century, Presbyterian missionaries engaged Ni-Vanuatu chiefs in debates about whose god was more powerful. For example, in 1880, John Paton decided that he would dig a hole to produce water as evidence that his knowledge and god were more powerful than any others. Similar stories abound in missionary records in other geographical contexts. Jean and John Comaroff (quoted in Asad 1996) write that the most significant issue in such debates is not whether or not the potential Christian believes the missionary, but the fact that what the missionaries were actually doing was compelling their would-be convertees to participate in European bourgeois forms of debate and evidence. Thus, Ni-Vanuatu have long been experts at understanding conversations about forms of knowledge and shaping them in ways that protect their interests. In the 1980s, certain people in Longana, on the island of Ambae, were "masters of tradition" (Rodman 1987) in adapting traditional knowledge to suit contemporary possibilities of how land was to be owned and used. McDonnell (2015, 2016) writes of powerful Ni-Vanuatu men who are able to negotiate the cultural, financial, and legal parameters of land dealings as "masters of modernity." Vanuatu's emergence as an independent nation was grounded in the incorporation of "human rights" and *kastom* knowledge into Ni-Vanuatu narratives of nationhood (Jolly 1996, 1997). I see quantifying life in the form of "alternative indicators of well-being for Melanesia" as another iteration of this adaptive syncretism. Recognizing that population-wide knowledge is necessary for evidence-based economic and social policymaking, "alternative indicators of well-being for Melanesia" speak to how lives can be well lived in this particular place and how collective traditional values can be protected.

How women's experiences and knowledge figure in articulations of traditional culture has been a troubling subject in Vanuatu, particularly in light of the ways in which certain global discourses are vernacularized. For example, Margaret Jolly (1987, 1996) has shown how the selective articulations of *kastom* have been part of the process of excluding women from economic and political leadership roles or from debates about their human rights. Important counter-discourses have been sustained on this topic – for example, through the development of the field-worker program for women at the Vanuatu Cultural Centre, a crucial space has been created for documenting and celebrating women's culture (Bolton 2003b). As well, "women's role in the traditional economy" has been documented through various projects at the Vanuatu Cultural Centre. This chapter aims to add to this conversation by showing how women's care knowledge and experience of care, especially during pregnancy and during the immediate post-partum period matters to the "alternative indicators of well-being."

Colonial officials and demographers have generated quantified figures to measure population health since at least the early twentieth century; as I have shown in the case of Vanuatu, these took the form of the "imbalanced sex ratio" in the 1910s and 1920s and "subsistence" in the 1960s. Though not formally called "indicators," these quantified figures made reproduction public and visible to colonial powers in ways that indicated the health of populations and economies, framing the objects and scope – if not success – of colonial planning, logistics, and intervention. Especially since the 1990s, indicators have become a key aspect of the collection and organization of data associated with health and social development, which has been made possible by the infrastructures of various international organizations (Merry 2016). In her ethnography of the world of indicators, Merry demonstrates how in the production of indicators, complex situations are rendered commensurate by categorizing and removing them from their social context. Issues are prioritized, ranked, and compared. Quantified indicators are seductive, she writes, for their truth claims that risk the homogenization and oversimplification of cultural and social structures (Merry 2016, 212–15). As indicators are constituted through the techniques and categories of data collection, they do not reveal a pre-existing reality; rather, they produce it (32). Indicators also contribute to governmental decisions by rendering knowledge of social processes public in particular ways. In short, quantifications of population health and indicators of well-being are not just descriptions of life, but, like a wide range of statistical representations of populations, are part of "social life itself" (Asad 2002, 78). The "Alternate Indicators of Well-Being for Melanesia" are occupied with quantifying the importance of social obligation, or "volunteerism", to both the economy and well-being, but, because of their economically oriented audiences and publics, can they account for the gendered care practices that are so crucial to reproduction?

Quantifying Population Knowledge, Melanesian-Style

In May 2010, at the Mele Wharf overlooking the black sand beach, I ran into Ralph Regenvanu, then a member of parliament and a former director of the Vanuatu Cultural Centre. While exchanging pleasantries, he mentioned that he was there finalizing plans for workshops on Melanesian well-being. Regenvanu has since founded a new party, Graon mo Jastis (Land and Justice), and has served as minister for lands, geology, mines, energy, and water resources, minister for foreign affairs; at the time of writing, he was the leader of the Opposition. He has been a leading advocate within the government for women's issues in Vanuatu. In 2016, when he was passing through Toronto on his way back from the University of Prince Edward Island, where he had attended a conference on climate change for small islands, he met with several people with ties to Vanuatu, mainly anthropologists, over lunch. The topic of research and reporting on "Melanesian well-being" came up again in that conversation, as something that he thought was necessary to counter the monetarization of development projects the World Bank often funds. As a country with a low GDP per capita, Vanuatu citizens are seen as needing loans, which then leads to land privatization in order to repay such loans, he told us.

Many Ni-Vanuatu who work in the NGO or government sectors are keenly aware of the importance of quantifiable, evidence-based measurements of well-being when engaging with development donors and making policy. In this context, the Malvatumauri National Council of Chiefs (MNCC) published the *Alternative Indicators of Well-Being for Melanesia* (2012). The Malvatumauri, composed of traditional chiefs (who are all men in Vanuatu), is an institution that aims to shepherd customary leadership and law within the country. The overall steering committee for the project included representatives of important institutions in Vanuatu and the wider region: the Vanuatu National Statistics Office (VNSO), the Vanuatu Cultural Centre, the Secretariat of the Pacific Community (the leading regional organization in policy research), the Melanesian Spearhead Group Secretariat (a regional organization that aims to promote economic growth), a member of the national parliament (Ralph Regenvanu), and the Asian Development Bank (an institution that aims to reduce poverty through loans). The Christensen Fund provided funding for the project. The 2012 results were the result of two years of fieldwork and analysis of focus group discussions and interviews from across the archipelago.

The express purpose of developing these indicators was to combat the trends in development assistance that Ni-Vanuatu have to contend with. In a 2010 interim report, the concept is described as follows:

> The almost universal use of GDP-based indicators to measure progress has helped justify policies based on rapid material progress at the expense of more holistic criteria. Because it [GDP] is a crude measure of only the cash value of activities or production, GDP is heavily biased towards increased production and consumption regardless of the necessity or desirability of such outputs. Policies developed with regard only to increasing per-capita GDP can have negative, and potentially disastrous, impacts on other factors contributing to life quality. (VNSO et al. 2010, 2)

Those involved in the project are keenly aware that employing numerical criteria to measure well-being can be at cross purposes, but they justify this procedure as follows:

> There are many who may be skeptical of an index that reduces well-being to a single number. For practical application, however, human well-being has to be translated into a metric system – without some kind of measurement system, well-being cannot guide practical policies and programs. If it is simply left at the level of inspirational discourse, conventional indicators will continue to play unwitting roles in Melanesian society. (5)

Quantifying Well-Being

The final *Alternative Indicators of Well-Being for Melanesia* report declares that Melanesian well-being includes "free access to land and natural resources, community vitality, family relationships, and culture." It defines culture as "language; sense of identity; core values, change in values and customs; status of traditional skill sets; access to traditional wealth; and participation in various cultural ceremonies" (MNCC 2012, 36). According to the report, beliefs must be accompanied by knowledge about how to convey them and pass them on. Consequently, the research team undertook an inventory of how widely skills deemed traditional are used. Ten traditional tasks were selected to stand for "culture": "mat weaving, basket weaving, broom making, thatch weaving (*natangura* or coconut [palm] leaf), wall weaving (bamboo or wild cane), tree planting (for use as posts, fencing, etc.), crop planting, food roasting, *laplap* baking (cooking with stones), and medicine producing" (43). Everyone who was interviewed possessed at least one of these practical skills.

Within that list, priority is given to five basic skills for "housing (mat weaving, wall weaving), feeding (crop planting, food roasting), and healing (medicine producing). Nearly two-thirds of those interviewed reported that they or someone in their household possessed all five basic traditional production skills" (MNCC 2012, 43). Traditional skills and participation in the traditional economy are thus extremely important, according to the report. These would of course be invisible if measured by indicators that focused purely on monetary economics.

The report deems Melanesian values and community vitality to be integral to the experience of well-being, because in Melanesia "society is not centered on the individual" (MNCC 2012, 73). Melanesian values include "honesty, strong family, going to church, respect for family, respect for Chiefs/leaders, helping others, respect for culture, working hard, reciprocity, trusting others" (73). The report also makes clear that these values are under threat and need to be nourished by a renewed focus on relationships. The research team found an increase in "perceptions of stronger selfishness and greed and dishonesty," a "consensus that cooperation has weakened in last 2–3 years," as has "respectfulness and prioritization of family" (74).

The national results were broken down by rural and urban differences, and then by province, before being reproduced in charts. A significant aspect of these numerical values is that they positively emphasize relationships rather than experiences of well-being located at the individual level. The report claims that participating in reciprocal relationships (the opposite of greed and selfishness) is what contributes to well-being.

The report's section on "community vitality" demonstrates that "social capital" is an important aspect of well-being. Social capital is measured through acts of "volunteerism" and "social obligation" for which no money is paid, including ceremonial exchanges and reciprocity of objects. Gardening (an activity in which 85.2 per cent of the national population participated at the time of the 2012 report) and childcare (in which 79.5 per cent of the national population participated) are important activities through which people give and receive unpaid work. Translating voluntary labour into social capital, the final report states that "it is clear from the information obtained that strong community networks are necessary for self-reliance – it takes more than a family unit to plant the food a family needs to survive. The level of voluntary childcare activities also attests to the high value of social networks in raising a family" (MNCC 2012, 60). The complex networks of child rearing and motherhood that I discussed in chapter 4 and elsewhere (Widmer 2010), the aspects of social reproduction that fall under "community vitality," are thus measured in terms of work for which cash payments were not exchanged but which nonetheless contribute to well-being and happiness.

VNSO's (2021a) *Well-Being in Vanuatu* report follows up on many of the alternative indicators in order to provide a robust baseline for the Vanuatu government's sustainable development plan. The report updates "community vitality" data from 2010, noting that 82 per cent of people over the age of fifteen in Vanuatu have someone who they can rely on when they are sick, while 77 per cent of people over fifteen have someone who will support them when they face financial trouble (18). Under the indicators of "reciprocity and exchange" (measured as a ratio of giving to receiving) the report notes the following:

> More than three quarters (78%) of the population aged over 15 give unpaid support to others in their community or area each year – 80% of males and 76% of females. This support comes in the form of time and labor to help prepare land for food crops, help build or repair household structures, prepare meals for community feasting, do work ordered by a Chief or community leader, care for another's children, or care for another's livestock. (18)

The authors of the 2021 report express some caution to readers and potential users of the data: "The act of giving and receiving unpaid voluntary support to or from another individual in one's community or area has decreased nationally since 2010 by 5% (drop in giving support) and 13% (drop in receiving support). This is another sign that the social safety net is weakening or getting smaller and leaving some people without the support they may need" (19). The section on reciprocity and exchange concludes as follows: "Thriving individuals have people they can count on in times of need. Cooperation and participation in ceremonial and community activities is an important contribution to the social safety net individuals must work to build and maintain, even when based in urban areas. It is all of our responsibility to give as much or more of ourselves as we receive from others" (19).

In documenting the indicators of well-being – part of the larger effort to create alternative indicators of progress that go beyond the standard measurement of GDP – the researchers frame well-being in relation to economistic, value-producing goals (although unpaid); this includes access to traditional material objects like mats and housing materials, and the knowledge necessary for utilizing the material objects in exchanges, or in daily life, and giving and receiving voluntary labour. Exceptions to the emphasis on forms of material exchange do exist, of course – for example, the indicators of knowledge of an Indigenous language and community participation in traditional decision making. The alternative indicators aim to quantify well-being in Melanesian terms, but, because such knowledge is to be used in the formation of social policy and development goals, many indicators are framed in relation to what are normally paid activities: the way the formal economy measures and assigns value sets the framework. The importance of care for well-being is mentioned and framed in terms of the people who can be relied on to provide volunteer labour while other qualitative aspects of care are under-reported.

Quantifying Women's Well-Being

The 2012 report indicates that because family and community vitality (key aspects of tradition) are thought to be stronger in rural areas, researchers also conducted a "Rural Community Well-Being Survey." This enabled them to perform a comparative analysis of rural and urban areas, which was important for their goal of documenting the "traditional economy."[4] Indicators related to women's lives appear in this section of the report. The data was collected by surveying women's leaders in rural villages. Researchers asked these leaders about "the conditions for women in their communities in regards to support, respect, and violence" (MNCC 2012, 89) to assess the larger context of women's well-being. In the summary of this portion of the report, the searing first point reads as follows: "Women's leaders perceive respect among men for women lowest for their own wives, followed by higher levels for their own mothers and highest for other men's wives" (xii). When questioned on the greatest challenge to well-being that women in their community now face, the most common response centred on spouse's behaviour. The report further explains:

> The main complaint regarding husbands was their lack of support and assistance with household duties, their kava abuse, overworking, unfaithfulness, and over controlling treatment. Lack of sufficient finances was the second most frequent response to the question, followed by gossip, violence, and uncooperative women. (89)

Another topic that the women's leaders were also asked to quantify was the number of women or girls "who became pregnant or gave birth out[side] of [a lasting] union in the previous 12-month period" (89). The responses indicate that pregnancies and births that occur outside of enduring unions are a common occurrence, and these were considered relevant within the report because of the "social implications of women having children out of union" (89). These are the village experiences of the "young mothers" I discussed in chapter 4. Furthermore, this phenomenon "can also serve as another indicator of respect for women at the community level and could serve as a support indicator of reproductive health" (89).

The concerns underlying the "alternative indicators of well-being" are similar to those that framed a set of projects that preceded them, the *kastom ekonomi* projects. These were developed by Ni-Vanuatu between the early 1990s and 2008 and emphasized their versions of "self-reliance" and "sustainability" in ways that would have been recognizable to "economists and finance people" (Regenvanu quoted in Rousseau and Taylor 2012, 181). Rousseau and Taylor (2012, 180) argue that "positioning self-reliance and sustainability as the alternative to Western consumerism, highlights the strongly materialist orientation of *kastom ekonomi* projects." Moreover, Rousseau and Taylor suggest that although self-reliance is framed as an alternative path to development, it does overlap with and replicate aspects of the neoliberal reforms undertaken in Vanuatu. What neoliberal reforms and the alternative indicators of well-being also share is that both privilege economistic thinking and thus underplay local practices of care by subsuming them under volunteerism and social obligation.

Alternative Indicators as Social Forms and Moral Objects

My intent here is not to critique the methods or intention of the alternative indicators. In fact, I stand in awe of their underlying intent as well as their outcome. Rather, I am interested in showing how the report represents a particular form of Ni-Vanuatu knowledge produced in the context of an attempt to live meaningfully through the latest iteration of capitalism. Like the preceding *kastom ekonomi* projects, it is an example of what Rousseau and Taylor insightfully describe as "the indigenous ... encompassing modernity in order to bring about change" (2012, 184). These indicators are aimed at creating public quantifications – tools for measuring the well-being and culture of a society that make non-monetizable values legible at a national scale in a way that is suitable for social policymaking.

"Well-being for Melanesia" works as a set of alternative indicators and a moral figure that quantifies and renders social reproduction visible. The metrics emphasize aspects of culture, mainly those associated with reciprocity and exchange (which have both moral and economic dimensions). As in other parts of this book, when reproduction and social reproduction are made visible and therefore amenable to policymaking, this is done in concert with both monetary and non-monetary economic concerns. These "alternative indicators of well-being for Melanesia" render Ni-Vanuatu social reproduction publicly visible by quantifying the non-monetizable value of reciprocity and making it legible for policy in terms like "social capital," "community vitality," and "volunteerism." They reflect the intent to protect non-monetized economies and values of collective well-being in ways that are legible to audiences interested in monetized economies. By creating the moral figure of "Melanesian well-being," Ni-Vanuatu are also attempting to use the language and form of contemporary governmentality to participate in international networks on their own terms. Ni-Vanuatu hope to assert their sovereignty by partaking in the collection of data within a quantitative framework intended to feature in conversations with representatives of external funding agencies who influence social policy.

Researchers have applauded the alternative indicators (e.g., Tanguay 2015), and they have since been taken up in policymaking. As reported on the Devpolicy blog, "the former prime minister of Vanuatu, Hon. Moana Carcasses Katokai Kalosil,[5] commenced national consultations in early 2013 towards a sustainable development plan, which will incorporate the alternative indicators in the monitoring of national policies, programs and projects beginning in 2016" (Tanguay 2014). When the *National Sustainable Development Plan* was published, as promised, the alternative indicators figured in the how the plan was to be monitored and evaluated (Government of Vanuatu 2017). The alternative indicators are present particularly in the goals of promoting Indigenous languages (14), improving "subjective well-being" as it relates to violence against women, children, and vulnerable groups (20), and strengthening links between traditional and formal justice systems (22).

In 2021, as I completed the final revisions to this book, the *Well-Being in Vanuatu: 2019–2020 National Sustainable Development Plan (NSDP) Baseline Survey* was published (VNSO 2021a). On 13 July 2021, Ralph Regenvanu posted a link to the survey on his public Facebook page, proclaiming that

> It is truly an historic milestone for Vanuatu to see well-being indicators mainstreamed into the activities of the Vanuatu National Statistics Office (VNSO). As we all know, using Gross Domestic Product (GDP) to measure Vanuatu's development progress is not realistic for a dual economy like Vanuatu because it cannot measure the "custom economy," which still remains the main economy in the country in terms of sustaining people's lives.
>
> Finally, after 40 years of Independence, we are starting to measure our development as a nation in terms of what exists in reality – and what we value – by including indicators that measure the "custom economy" and incorporate "Melanesian values" into the national policy arena.
>
> The "Well-being In Vanuatu" report is the work of several years of planning, fieldwork, data processing and analysis following the 2016 Mini Census that attempts to look at well-being more holistically, framing it around topics of happiness, access, knowledge, health, and social resilience. Some of the indicators developed in the "Alternative Indicators of Well-Being for Melanesia" pilot study from 2010–2012 have become key monitoring and evaluation indicators for the National Sustainable Development Plan (NSDP), securing their collection through to 2030.
>
> Many indicators that are now prominent in the Vanuatu development framework are relatable across the Pacific. If the work Vanuatu is doing can help influence other countries in the Pacific to be more concerned with the well-being of their people rather than focused on endless GDP growth, that is the future we want to see. A future with balanced and well-being centred governance and development. (Regenvanu 2021)

It remains to be seen how this knowledge will actually be implemented and what its effects will be. What interests me here, though, and what is relevant for understanding knowledge about well-being and social reproduction in Vanuatu, is the fact that this knowledge was asserted at all. Moreover, I am interested in the way that this knowledge casts itself as a measure of health ("well-being"), adopting a crucial political idiom as well as a powerful way of articulating identity, cultural assertions, and histories of injustice (Adelson 2000). The report on alternative indicators is thus a socially significant form of knowledge that is asserted to global audiences and is concerned with how lives can be lived well in the face of the current iteration of capitalist transformation.

When health and well-being in particular are made into quantified figures – a phenomenon now so central to the organization of contemporary global health – the resulting metrics shape what gets seen as a priority health issue and therefore resourced by governments, NGOs, and international organizations (Adams 2016a). In this sense, the figures are political as they shape what becomes visible as the object of intervention, support, and governance. In addition to being political figures, the alternative indicators are moral figures insofar as they are concerned with demarcating a specifically Melanesian development path. They also assert sovereignty through the collection of data, though the quantified framework is in place to speak to social policymakers and external funding agencies and donors who value GDP-related indicators.

To return to a core focus of this book – namely, the importance of care practices and knowledge for reproduction in relation to quantification – I now look at an aspect of well-being that is downplayed in the indicators. In particular, I look at forms of care given and received by women during pregnancy that go unmentioned in the national-scale alternative indicators but that is central to pregnant women's well-being. By doing so, it may seem as though I am presenting a variation of the familiar argument – made especially by feminist economists – that scholars should make women's care visible because of what it contributes to the economy or public sphere (e.g., Folbre 2002, 2008; Waring 1988). Engaging in critical conversations with such scholarship, feminist anthropologists have recognized that care provided within kinship networks is central to the workings of capitalism (Bear 2013; Shever 2013), as well as the social reproduction of class, gender, and racial inequalities (Rapp 1992). Consequently, feminist anthropologists have argued that it does not make sense to consider "the economy" or "the public" as bounded domains – indeed, they argue that to see the "productive" sphere as separate from "reproduction" is merely to reinforce the ideology of capitalism. Rather, such critiques argue, the domains of "the economy" or "the public" should be seen as co-constituted by historical and social circumstances (Bear et al. 2015; McKinnon and Canell 2013). In this chapter, I show that the intended public, or scale, addressed by the alternative indicators means that reproductive care is rendered invisible not only by capitalist forms of exchange (as has long been shown), but also in alternative indicators that focus on non-monetary economies that themselves aim to protect Indigenous economies from capitalism.

Massage healers are moral figures who deliver a type of care that cannot be quantified. They are moral figures because their knowledge of well-being is concerned with moral conduct, especially with the need to behave in a reciprocal way. The alternative indicators and the massage healers' expertise both attempt to connect well-being to reciprocal relationships, but the particular kind of care provided by women massage healers, which is crucial to women's well-being, especially during pregnancy, does not feature prominently in the alternative indicators of well-being.

In a chapter about the quantified figure of "well-being for Melanesia" and localized expertise in pregnancy care, these forms of care and authoritative knowledge could be perceived as another iteration of the "politics of *kastom*" for which Pacific Islanders are well known. Indeed, *kastom*, a selective iteration and protection of traditional culture, has been part of many aspects of life in postcolonial Vanuatu, from assertions of nationhood (Tonkinson 1997) to the protection of intellectual property (Geismar 2013) to development of the tourism industry (Cheer, Reeves, and Laing 2013). The report on alternative indicators, being directed at an international audience, does not limit the discussion to *kastom*; it also uses the language of "Indigenous customary land" and "traditional" knowledge and exchange. In VNSO's (2021a) report, *Well-Being in Vanuatu*, *kastom* is also mentioned in reference to stories, dance, songs, games, reciprocity, and land. And yet, massage is not commonly referred to as a *kastom* practice. *Kastom* can also refer to sorcery in the context of illness, but sorcery is never part of discussions of *kastom* that circulate for national or international audiences. Moreover, as I will discuss, these healers only recognize sorcery as a possible causal factor and would not want to be perceived as sorcery experts.

Creating Female Spaces of Care

Touser, a well-loved grandmother in Pango Village, was just finishing massaging two children when I arrived at her orderly corrugated iron house. She is part of the family I know best in the village, and we had arranged beforehand for me to interview her. The children slept on small pandanus mats while their mothers looked on. As well as massaging them, Touser had given them a medicine to drink, so that they would *kam gud* (get well). Touser invited me in along with Daniella and her female relative, who had arrived at the same time. Daniella, who was about seven months pregnant, had just married a man in the village. My children, partner, and I had been part of the throngs who had attended their wedding. The preparations for the festivities had occupied many in the village for weeks, creating spaces of conviviality at a time that was otherwise quite tense because of recent land sales and a dispute over chiefly title.

I offered to leave so that the treatment could take place without potential distraction. However, Touser encouraged me to stay. "*Yumi ol woman,*" Touser said, using the inclusive Bislama pronoun (*yumi*) for "we," which clearly included me in her statement. This word contrasts with *mifala,* which would have excluded me from the "we" to which the speaker referred, and the overall phrasing indicated the feminine orientation of both the space and the care that was to take place. Touser began preparations for Daniella's massage by placing a pillow, a small mattress, and a pandanus mat and sheet at the far side of the room (which was also part of Touser's home). Daniella then lay down on her side, continuing to wear her spacious island dress that afforded her modesty during the treatment. Touser poured water from a bottle into a plastic basin and dipped her hands in the water. She held them below Daniella's pregnant belly and gently lifted it up and down. Touser would later tell me the cold water wakes the baby up, enabling her to move it gently into the proper position.

We sat in companionate silence for about fifteen minutes while Touser did her work. The one-room house was dimly lit by shafts of sunlight entering from two window openings. A table and chairs were placed in one part of the room, while the opposite end featured an open space with a pile of pandanus mats. Hand-woven baskets, used to carry food and other household goods home from the garden, as well as providing storage for these items, hung from the rafters. Touser gently talked to Daniella. She then took oil from an unlabelled glass jar and rubbed it into Daniella's back. Her hands moved from the bottom up in long, gentle strokes and then lingered in certain places. Touser then moved Daniella onto her other side and repeated this sequence of actions. She moved her back to the first side and once again repeated the process. Touser turned to ask Daniella's family member if she had fallen down. This individual replied that she had not, but that she had been in a car accident. Touser would later tell me how proud she was that she had been able to feel the accident in Daniella's body. Daniella looked relaxed and peaceful afterward, commenting that she could not breathe deeply before the massage, but now she could. The baby had

had too much partying at the wedding, Touser gently joked, and it was suffering because Daniella had been standing up too much. Daniella's relative got up and they left together, handing Touser a small bag of rice, a common exchange object.

At this time, I had been hearing in other interviews and conversations about the exceedingly common occurrence of intimate partner violence, which many women face and have faced for three generations, and other dimensions of violence against women (a well-documented problem in the south-western Pacific region; see Eves 2019; McKelvie et al. 2020; Wardlow 2020). A phrase that I had frequently heard from women of all ages was "life is hard in Vanuatu." Against that background of gendered hardship, the care and peace that this female space offered took on a heightened importance. All of the pregnant women I spoke with would visit such healers at least once – and often several more times – during their pregnancies. Given the dearth of biomedical resources devoted to pregnancy care, and the conventional indicators of women's and reproductive health discussed in chapter 4, the environment that the massage experts create is one of the few significant spaces of care and well-being available to women.[6]

There are several spaces of female sociality in Pango. For example, there are various women-operated small businesses selling prepared food in the village, or handicrafts at the wharf to passengers of the immense cruise ships that dock in Port Vila. A more recent and particularly relevant example is Mama's Laef, a business developed around sewing and selling reusable menstrual pads. For her leadership role in this enterprise, Mary Kalsrap has been recognized as one of "70 Inspiring Pacific Women" (Pacific Community, n.d.). Beyond these enterprises, there are other spaces of female sociality. In the late afternoons until it gets dark, young women play netball at the school or on the village field. Older women work together on church projects, or garden together on their pastor's land. Within this constellation of female social spaces, the spaces of care that the massage healers create are central resources for the nurturing of women's well-being during pregnancy and for the care of small children.

Massage is a form of care that is far more commonly used for well-being than biomedical care during pregnancy. An overwhelming sense that the hospital is incapable of offering sufficient care pervades young women's stories of childbirth. As I showed in chapter 2, this feeling about births in the hospital is not shared by older women who gave birth at the mission hospital in the 1950s and 1960s. This is in spite of the fact that the existing technical infrastructure at the hospital, while far from what it should be, is more substantial now than that it was in the 1960s. Like the people in Madang, Papua New Guinea (Street 2014), the women in Pango spoke about the lack of medical infrastructure as evidence of the government's lack of care.

Massage is an important part of pre- and postnatal care in many parts of the world for enhancing the well-being of the pregnant woman and the fetus (e.g., Kildea and Wardaguga 2009; Davis-Floyd 2018, 234; Jordan 1993, 26; Cosminsky 2016, 73–7). Massage techniques have been shown to entail embodied and intuitive knowledge that women seek in order to relieve discomfort, position the fetus as advantageously as possible, and heal the uterus after birth. The practitioners are generally known to the women, very limited technology is needed, and the expertise is financially accessible. These points hold for the healers I spoke with. The healers also create spaces of care.

Massage healers' knowledge relating to well-being and reproduction involves an authoritative assertion of knowledge of reciprocal social relations – capitalist and otherwise. The pioneering anthropologist of reproduction Brigitte Jordan (1997, 56–7) writes that authoritative knowledge is persuasive because it seems reasonable and effective in a particular context. A knowledge system can become authoritative if it can "explain the state of the world better than others" and/or have a stronger institutional power base (55). Biomedical knowledge is often considered a type of authoritative knowledge (e.g., Liamputtong and Kitisriworapan 2014), but alternative knowledge can attain this status through certification (e.g., Henley 2015), while embodied and intuitive knowledge (Teman 2003; Davis-Floyd and Davis 2018) can also become authoritative. The massage healers do not directly compete for authority with the biomedical system; indeed, many young women wish for more biomedical care and greater access to biomedical knowledge concerning fertility control.

Healers' Skills

I spoke with six of the most well-known female massage healers in Pango. The women's assertions of authority emerged in my interviews with them after I asked about their work, rather than from my own observations of them asserting their claims in daily life. Most of them belonged to the oldest generation of their families. Two had been trained as nurse-midwives at the mission hospital in the 1950s and 1960s. They were from different Christian denominations, ranging from Presbyterian to Pentecostal to Jehovah's Witnesses. In Bislama, the verb for what they do is *holem*. This word means to caress or hold as well as to grab or hold on to an object or to maintain or preserve something. When I met with these women, we spoke about the types of problems people bring to them as particular experts in their village, and we focused on their skills and knowledge, particularly those associated with pregnancy and reproduction.

Before becoming pregnant, a woman might seek out someone to help them. The central reason for this, as stated by the women asking for care, is to ease their discomfort (it is worth noting that women who are not pregnant or trying to get pregnant also seek the services of a massage healer for the same reason). Treating infertility or the unexplained cessation of menstruation is also a reason for a woman to visit a massage expert before her pregnancy. Expert massage healers have common ways of explaining miscarriages and various infertility issues, how they would resolve these problems, and why massage is helpful. They say that the massage makes the blood flow properly again, because *rabis blad istap* (there is bad blood). The massage "puts the uterus back properly."

The healers commonly explain that they massage a pregnant woman to make her breath "go properly," which entails gently moving the fetus, as Touser had done for Daniella. Women would visit a healer for a variety of other reasons as well. Sometimes they needed to recover from a vehicle-related accident or an accident at home, but most commonly, they seek help for pain that is said to be caused by too much work: lugging and washing clothing by hand, carrying children, getting food from the garden or shops in town, cooking and generating some cash income from a small businesses – really any of the various activities associated with keeping a household going. The healers would talk about women's work, as would many of the women themselves, as requiring significantly more time and effort than the work of men. At times, as some healers told me, women would come and just cry. Thus, the care the massage healers provide is for women who are weary from the work of making relationships within families work, and weary of the violence that they or their friends and kin are experiencing. The latent or direct presence of violence could weigh heavily on their bodies and emotions.

As I have described, massage care is needed to attend to the realities of unequal work, as well as the injuries caused by violence or various types of accidents. Some of the women massage healers I spoke to also had knowledge of leaf medicines. As I noted in chapter 2, this knowledge tends to be passed down from mother to daughter, in line with the common practice of keeping knowledge within matrilineal groupings. Massage expertise – knowing how to touch and put body parts back where they belong – is also perceived by some Ni-Vanuatu as a "talent" emanating from God. Until the mid-twentieth century, women in Pango with these skills would normally have tended to women during labour and the immediate post-partum period (as discussed in chapter 2). However, it is becoming difficult for the healers to pass on their abilities, as the current generation of young women (who are the ones needing the care mentioned above) tend to show little interest in learning their mother's skills. This follows a trend in Vanuatu, commonly remarked upon, whereby the importance of cash and wage labour in people's daily experiences has led young people, especially in peri-urban places, to show more interest in town-based jobs than in learning skills from their parents' or grandparents' generations. This is also part of the longer-term trend towards urbanization and the expansion of capitalist forms of labour that I presented in chapter 2.

Since at least the 1960s, people in Pango have contributed in many ways to the colonial economy through their labour as police officers, telegraph operators, nurses, doctors, domestic workers, and retail shop workers. Older people recount that their wages then were often too low to sustain them, so gardening remained important. Today, with the expansion of the service economy, more women are working for wages as domestic workers and in the civil service, health professions, the banking industry, and at NGOs. There has been a decline in garden work connected with the widespread presence of resorts occupying the land that had previously been a significant source of shelter-building materials and food. Though no longer central, gardens are still important as places for food production, companionship, and peace and quiet. Men can work as clergy, health professionals, in the civil service, as resort staff, and as taxi or bus drivers; they may also work sporadically in construction or gardening. Households with limited access to wage labour hold "fundraisers" at which they sell cooked food and produce from their gardens or more precious forms of produce sent via kin or church networks from other islands. All of these wage-earning possibilities are part and parcel of a peri-urban context in which wage labour is increasingly necessary but also time-consuming, and which in that sense is different from the rural islands that comprise the majority of the country. Women's common perception of men not working is a source of tension within families. Given this context, young women do not show a lot of interest in learning the healers' expertise. However, even though they themselves may not be interested in learning to practise massage, young women do very much recognize its importance for their own well-being.

Well-Being and Knowledge of Reciprocal Relationships between People

Mothers often bring their babies as well as their preschool children to these female massage healers. Children who would cry persistently without reason, or who stopped nursing without reason, would be massaged. If the baby did not get better after a certain amount of time, the healer would tell the mother to visit the hospital or visit a *kleva* (an expert in sorcery remediation). On one occasion, I sat with Martha at a wooden table in the shade of her corrugated iron and bamboo shelter next to her small store and living quarters. A married woman with grown children, Martha told me that to stay well, pregnant women have to protect their chests from cold and rain (cold in this instance means below twenty degrees Celsius) and avoid walking around at night. This last piece of advice – not walking at night – was often aimed at protecting someone from sorcery. Relatedly, women were advised to avoid hanging the clothes of their infants (up to six or seven months) on laundry lines at night, the reason being that *nakaimas* (sorcery) could cling to the children's clothes by mistake.

Martha proceeded to tell me about a family on another island in Vanuatu that had lots of children and so allowed one child to be taken to Pango by a family member to live with a Pango family. The child got sick, and the doctor could not help. Martha said she recognized that this was a problem of *nakaimas*, and so she massaged the baby's legs (which were unusually crooked, she said, like a snake). When the baby left, Martha and her husband prayed over the child, just as they pray over all of the sick people who come to see them. I expressed surprise that anyone would want to harm a child in this way. The problem was, Martha told me, that the family on the other island probably sent *nakaimas* to the baby because they wanted their child to come back to their island. The family in Pango did not adopt the baby properly, which entails giving a pig to the child's natal family. Such exchanges are important life-cycle events and as such play a key role in social reproduction. Martha demonstrated her authority by recognizing the potential for sorcery because the proper reciprocal exchanges had not been made, and, therefore, relationships had not been made right.

Well-Being and Knowledge of Reciprocal Relationships between Humans and Non-Humans

As a young adult, Susan, another female healer with adult children and small grandchildren, had lived with an expatriate medical practitioner and his wife. She had been intrigued by the doctor's medical books and the photographs of medical problems that he had shown her on many occasions. These images, she said, taught her how muscles could be out of place. She now lives from the produce of her garden and makes some cash income from her sewing machine, her prized possession. Susan's stories extend beyond the treatment of pregnant women and children and provide a broader context for these forms of reproductive expertise and care. In addition to learning from medical textbooks, Susan's mother-in-law also taught her how to do massage, but she never applied this skill until one fateful day when her mother had a stroke. Susan took her mother to a *munwe*, a female expert in healing sorcery; she was from the island of Tongoa but was then living in Seaside, an area in Port Vila. However, at that time, before the widespread use of buses and trucks, the distance between Pango and Seaside was too far for Susan to come and go as frequently as needed to continue the treatment. So, the *munwe* from Seaside blew on Susan's hands, and that gave her the power to heal. Susan massaged her mother, and her mother slept well.

When a small child close to her home got sick, Susan went back to the old woman in Seaside. When told what had happened, the old woman laughed and said, "You didn't compensate the spirits. They left the old woman and went into the child." So, Susan put 100 vatu (about CAD$1.15) in each of the old woman's hands for the spirits. And that is how she acquired her talent. Talents, Susan told me, must be used "properly from your head to heart to hands." Since helping her mother and the sick child, Susan has done just that. She frequently massages pregnant women, using oil to expertly caress swelling bellies and, if possible, to shift the positions of the fetuses. Until the woman is four to five months pregnant, Susan claims that she is able to turn babies into a favourable position for birth and tries to make the fetus sleep well.

A gifted and enthusiastic storyteller, Susan insisted on telling me another story that provides context for her expertise.[7] Once, there were a man and a woman who lived on Erakor Island (a small island near Pango where no one lives anymore because the village on Erakor was destroyed by a tidal wave and cyclone). They had one daughter and no other children. They went down to the beach and the girl played. In the sand, there was a mound. They did not realize that the ancestral spirit who lived in the mound was destroying their house. Appeasing the spirit meant offering something – namely, their girl in marriage. So they prepared mats and made food for the exchange ceremony.

Later, while the family was seated in their canoe in an area where there was normally very quiet water – Susan emphasized this point – a wave came in and the canoe tipped over. Then a snake rose up from the hole and took the girl. Since then, all sea snakes are the offspring of their union. The sea snakes are the *natatok*, the true custodians of the place. Susan can communicate with the snakes because she is from the same *naflak* as the people whose daughter was given in marriage. This enabled her to communicate on behalf of others with other *naflak*, who have unexplained bodily ailments that are untreatable at the hospital. For example, a man had a swollen body that doctors could not fix. After diving day after day, he found a coconut crab in a tree trunk at the shore near a black and white snake (an unusual occurrence). So, Susan went down to communicate with the snake and determined that no exchanges were needed; just talking solved that problem. She could communicate with the snake in ways that the man's family could not, because they are from a different land-based yam *naflak*.

In another story, Susan recounted how she treated Gladys, a woman in the village who had built her house on the *kastom ples* (customary place) of the sea snake, the *natatok*. Gladys and her family did not show due respect to this *natatok*, and they even built a table right on top of its *kastom ples*. Gladys should have offered the snake some food to show respect. Instead, she killed it under the table. Soon after, Gladys got sick, so Susan went to visit her in the hospital. Gladys could not even talk, and the doctor said that she was about to die. Susan went to the shore where the man and woman had given their daughter in marriage to the sea snake. She proceeded to talk to the snake in the hole. "Please give back Gladys," she implored. Then there was a wave in a place where there was normally perfectly calm water. Seeing this, Susan threw 100 vatu towards the *natatok*. This was compensation because Gladys had killed the snake. Susan made it clear that the 100 vatu was not a payment; it was compensation. Otherwise, the *natatok* would not be happy. Susan asserted her expertise through her ability to recognize unusual aspects in the landscape, like uncommon water patterns or the presence of snakes, that require intensely local knowledge. As well, she asserted her expertise and authority by knowing how to engage in reciprocal relationships.

At a moment when resorts are increasingly changing the landscape and reducing Ni-Vanuatu access to the sea, I wondered how this might be affecting the *natatok* and people's relationships with them. Given the tensions around land sales, this was not an easy topic for me to broach. People whom I knew quite well told me, "the *natatok* are the true custodians of the place, and it [whatever place was now occupied by a resort or for the most part could be seen or accessed only by those staying at the resort] remains their place." Otherwise, the response that I received was *yumi no save* (we do not know).

Economies of Care

All of the massage healers said that they did not have set fees for their care. Indeed, they all said that people should give according to *ting ting blo olgeta nomo* (according to their opinion). While this may sound like "pay what you can" when considered in a Western context – and to some extent it is – it would be more accurate to say that people give according to what they think this relationship means. Certain items would be remarked on – "Mary gave me an island dress" – as a particularly impressive gift, distinct from the more common gift of a bag of rice or sugar. Sometimes the healers begrudged a lack of generosity on the part of the women who visited them. But all of the healers have other income sources (e.g., sewing or operating a small store selling tins and snacks), or are cared for by younger members of their families. They do not see their expertise as being in conflict with biomedical care; they expressed the wish that the hospital could keep the women longer after birth, and all of them wished that the hospital in town was better or that there was a medical clinic in the village. Their expertise is intensely local and based on knowledge of the landscape, the morally appropriate amount of work that people should do before they get sore, and the right kind of reciprocal exchange.

Conclusion

I began this chapter by recounting my conversation with Joseph, in which he expressed his fear that people would be begging in the street because of the dramatic changes that had recently occurred in Pango. Several resorts had been built in the village, blocking visual and physical access to the open Pacific Ocean for many residents. Prior to the construction of one resort in particular, people in the area could welcome the cool breeze off the open ocean in this spot that marked a transition between village and town. The last golden slices of the setting sun were visible over the breakers crashing on the nearby coral reef. With visible changes like this in the village, and with accompanying dependence on too few waged jobs, Joseph, like many in Pango, worried about the community's well-being in the future. The possibility that they might someday have to beg on the street, a stark example of a non-reciprocal relationship, crystallized their concerns about how life would continue.

As we saw, one proactive national response can be witnessed in the formulation of the indicator of "well-being for Melanesia," which aims to quantify the importance of those Melanesian values and socialities that exist outside of commodified exchanges. "Well-being for Melanesia" is a moral figure for the value it places on good reciprocal relationships. In this sense, the statistical quantification required to make social reproduction public serves not only as an external technology of governance, but also as a means of critiquing the kinds of economies associated with contemporary development, as well as a tool for asserting sovereignty. Well-being, as depicted by the alternative indicators, thus becomes an Indigenous critique of overly monetized relationships and their impact on well-being. In so doing, it expands our thinking about the significance of metrics, indicators, and well-being itself.

The moral figure of the female massage healer shows Ni-Vanuatu knowledge that, like the quantified knowledge in the indicators used for policymaking, connects well-being to the value of reciprocity. These healers know that the pressing issues of women being overworked and subject to a myriad of worries lead women to visit the massage healers to treat the pain in their bodies. Massage healers assert their knowledge through an intensely local understanding of the landscape and a sensitivity to reciprocity in exchanges among humans, and between humans and non-humans. This sensitivity to the reciprocal balance in their environment also extends to the care they offer the community, in the process maintaining a vital space of refuge, care, and belonging for women.

A reading of these two moral figures and associated scales of knowledge of well-being reveals the importance of care that is not made public in quantified indicators. These spaces of care are vitally important for women's well-being, as are the exchanges between humans and non-humans that constitute economies of care. To honour the importance of care, particularly women's care, which is often taken for granted, feminist social scientists have often sought to demonstrate its importance as a form of labour that may be "unpaid," "reproductive," or "emotional." These analyses stem from a concern with demonstrating the value (often monetary) of unpaid labour and its role in the workings of capitalism, and to show how care work in the form of "good customer service" is integral to post-Fordist economies (e.g., Hochschild 2012). In places like Vanuatu, which have never been "Fordist" or "post-Fordist," and where most exchanges are not monetized, illuminating the importance of care, as I have tried to do here, assumes a different salience. Ni-Vanuatu articulations of the importance of care and reciprocal relationships have more to teach us than the importance of unpaid care for the monetary economy. By juxtaposing the two scales of knowledge foregrounded in this chapter, I have shown that the Ni-Vanuatu articulations of reciprocity are at once a critique of monetized social forms and a cautionary reminder of the ways in which relationality and care practices can be occluded. Care is made opaque in the alternative articulations of quantified well-being explored here, which themselves are designed to speak back to monetized social relations. Because the healers' assertions of well-being also focus on righting non-reciprocal relationships beyond capitalism, I have placed their knowledge in conversation with the assertions of the alternative indicators in an attempt to show the importance of women's care work and knowledge for understanding well-being, as well as the implicit critique of non-reciprocal social forms that it contains. The different scales at which this knowledge is communicated reflect a diversity of Ni-Vanuatu engagements with capitalism and conventional development agendas. The knowledge and care practices demonstrate a profound ingenuity on the part of Ni-Vanuatu when it comes to defending these values.

Epilogue

Relations of Reproduction and Survival in the Anthropocene

Life remains elusive, in excess of the infrastructures of its management and valuation.
Michelle Murphy (2017, 141)

Many will remember the first months of 2020. Across the world, people in different situations faced weeks of social isolation, physical distancing, job losses, premature death, changes in hospital protocols, frustrating searches for reliable information, increasing workplace risks, and desperation arising from the changes in their lives. Anxieties relating to many dimensions of life were pervasive. During this time, I was overcome with a renewed sense of what it must have been like for people living in Vanuatu in the late nineteenth and early twentieth centuries whose villages were radically reduced in size. Infectious diseases introduced from outside may have taken between 50 and 90 per cent of the population of certain villages. Ni-Vanuatu worked at maintaining, rebuilding, and reimagining the social fabric of their societies during *decades* of epidemics. The questions posed by early twentieth-century European researchers about why, after decades of watching their children die, women wanted fewer children, seemed newly and desperately ignorant and devoid of empathy for what it meant to face the unknowns of infectious disease epidemics.

In February and March 2020, like many people around the world, I read reports of grandmothers and grandfathers dying in northern Italy and of African American and Latino deaths skyrocketing in the United States. There was a widespread fear of economies collapsing, and then there was news of governments – who for two decades had claimed to have no money for health and social programs – beginning to bail out airline industries and, in the case of Canada, recommitting to supporting the construction of petroleum pipelines. There was concern about what "the new normal" might look like. Urgent questions were being asked: What are the costs to privacy if governments mandate the tracking of people's movements through cell phones? How and why does science matter? What is "essential work?" As I scrolled through my social media feeds, I experienced bouts of rage directed against at least two decades of neoliberalism, which have not only resulted in cutbacks to infrastructure and social services but have also impacted how individuals perceive themselves to be part of collective responses.

This is not the epilogue I had imagined I would write. With most of this book in manuscript form, I had planned to return to Vanuatu in March 2020 for three months to engage in follow-up conversations with the people whom I had interviewed in 2010. I had hoped to see the people I had come to know there: the retired nurses, "young mothers," and massage healers. I wanted to see if and how the discourse and concern about rapid population growth had changed. In particular, I wanted to see how concern with climate change was now entangled with notions of reproductive choice. And I wanted to know what changes the care economies involved in raising children had undergone.

In late February and the first half of March 2020, I spent many anxiety-filled hours following the news and the changing border regulations, considering whether I could still return to Vanuatu with my partner and two daughters, now aged eleven and fourteen years. On 13 March, while in Nelson, New Zealand, I decided that the risks were too great for all concerned. I could not risk bringing the virus to the communities where I would do research. My family and I drove to Christchurch and from there flew back to Auckland (where I had been doing archival research in February), and ultimately to Toronto. We passed through the spectacular landscapes of New Zealand's South Island with heavy hearts. The Malborough region of New Zealand, with the grapes nearly ready to be picked, seemed deserted. This is a region where many Ni-Vanuatu obtain work through a labour mobility program known as the Recognized Seasonal Employer (RSE) scheme. They generally work in the horticultural industry, producing food for New Zealanders to consume and export.[1] Throughout this short time on the South Island of New Zealand, I looked at the green hills and thought of Katerina Teaiwa's analysis of how these lands were made fertile by the phosphate sourced from Pacific Islands. She shows, devastatingly, how the agricultural and labour circuits that made New Zealand so fertile have made other Pacific places uninhabitable (Teaiwa 2015).

While in Nelson, I met, at a two-metre distance, with a consultant from Mama's Laef, the social enterprise involving people in Pango that I mentioned in the last chapter. They make and distribute reusable menstrual products and diapers. I had offered to bring a small package of materials to Pango. I learned that Mama's Laef began in the aftermath of Cyclone Pam. I wondered at that moment what new initiatives would come from managing the border closures that would inevitably follow. By that time, I already knew that I would have to return to Toronto, and so learning about the work that the consultant was doing with Mama's Laef and hearing about people whom we know in common was deeply bittersweet.

On 23 March 2020, Vanuatu closed its borders to all international travellers except for essential personnel. This decision followed in the wake of conversations about whether to close the border on social media: some worried about the devastating impacts that such a closure would have on tourism, while certain doctors begged the government to close the borders because the medical system would collapse if the virus became widespread in Vanuatu. The RSE was also suspended. From social media, I learned that Dr. Basil Leodoro, a Ni-Vanuatu doctor, was leading handwashing demonstrations in villages. The Kam Pusem Hed (a drop-in clinic for family planning and reproductive health) posted a notification on Facebook on 25 March that women who had appointments in April or May for family planning pills or injections should come in as soon as possible to obtain a three-month supply; the notification explained that if a COVID-19 case was announced, the clinic would close.

Though there were no COVID-19 cases, a state of emergency was declared in Vanuatu on 27 March 2020. Businesses were required to set up handwashing stations. These stations, depicted in photographs on social media, consisted of large plastic bottles with spigots, positioned at the edge of a table along with soap. *Nakamals*, extremely popular social venues where kava is consumed, were restricted to providing takeaway services. The Vanuatu government issued orders to contain the spread of misinformation about COVID-19 on social media: anything related to the virus needed the approval of the Ministry of Health before being posted. Prominent politician Ralph Regenvanu encouraged people to stay home and plant food. During the negotiations that followed the election held on 19 March 19 2020, Regenvanu emerged as the leader of the Opposition on 20 April; Bob Loughman was now the new prime minister.

Back in Toronto, a friend dealing with an enormous amount of labour to ensure pandemic preparedness in her already extremely busy job texted me to say that she never wanted to hear the word "unprecedented" again. Indeed, the word justified a vast range of social arrangements with implications for labour, education, mental health, and the separation of home and work. She, I, and many others were witnessing what to us were "unprecedented" developments. Being able to plan, as she and I both know, is a privilege, and while uncertainty of this kind is "unprecedented" for us, its meaning and referents are profoundly historically and socially contingent.

On 5–6 April 2020, the category 5 Cyclone Harold crashed through Vanuatu. When it had passed, it left over 160,000 people in dire need of assistance, having lost their homes, churches, schools, and food and water supplies. Heart-rending stories of survival emerged in the media. Radio New Zealand (RNZ) reported that a group of about sixty people on Pentecost Island "hid under the church and it flew away. All the other houses had flown so the men stood around in a circle and the women and children stood in the middle. They stood there until the morning in the rain and the wind" (Robinson-Drawbridge 2020, n.p.).

Mr. Warri, who now lives in Port Vila, conveyed the scale of the damage in his home community of Pentecost Island to an RNZ reporter. "The whole place looks as if it had been bombed," he said. "Our community is remote. It's up in the hills, so yesterday we chartered a helicopter to go over there and today we put things on a charter flight because we can't wait for the government" (Robinson-Drawbridge 2020, n.p.).

The Vanuatu government authorized surveillance of the damage from the air, when planes were finally able to fly over the area. The National Disaster Management Office (NDMO) needed to confirm on social media that the first airplanes that flew over the islands were assessing the situation and would not be landing. People clearly remembered that in the wake of the category 5 Cyclone Pam, which destroyed 90 per cent of the buildings and infrastructure in the country in March 2015, an army of humanitarian workers converged on the islands from Australia, New Zealand, China, and other countries. Such workers, the Vanuatu public was assured, would not be coming this time; the border would remain closed. If any workers did come, they would be quarantined for fourteen days. The NDMO posted images on Facebook of relief supplies arriving from China, Australia, and New Zealand and being disinfected on the runway to allay fears of the COVID-19 virus penetrating the country. Anthropologist friends and colleagues discussed on social media how to contribute money to support people and communities who had been so generous to us, both directly and indirectly. Like many overseas researchers, I *kivhan* (give a hand, help) as I can.

Jonathan Pryke, the director of the Pacific Islands Program at the Lowry Institute, an Australian think tank, told the BBC that "the economic impact of the cyclone on top of the economic fallout of Covid-19 is the last thing these countries need. Already stretched government resources will be stretched even further." However, he expressed his faith that the region would recover, saying, "the Pacific peoples are very resilient. They will persevere through this" (Tan 2020, n.p.).

Amid the humanitarian response to Cyclone Harold, which took place during the period of closed borders and concern regarding the novel coronavirus, the *Vanuatu Daily Post* (2020b) reported some good news on 18 April. The Pacific Private Bank (PPB) had delivered on a 2017 promise to supplement funds donated by the Australian government for the new maternity ward at Vila Central Hospital with a sum of 7.2 million vatu (approximately CAD$83,000). As described on its website, PPB is "an international bank owned by European entrepreneurs headquartered in Vanuatu for more than 23 years. It is dedicated to servicing international clients and companies from around the world and offers international transfers, term deposits and investment services" (PPB, n.d.). It is based in Vanuatu because the country's "geographical, economic, tax, and legal environment is favourable to protect capital for foreign investors" (PPB, n.d.). This donation to the maternity ward will enable further improvements following the substantial renovation work that has already been paid for by the Australian Department of Foreign Affairs and Trade as part of the Cyclone Pam Recovery Fund. During this construction work, the *Vanuatu Daily Post* reported that "the maternity ward's approximately VT30 million [approximately CAD$350,000] facelift is a godsend for all expecting mothers who have voiced their concerns of the dreaded original condition of the birthing ward" (Tokona 2020, n.p.). The number of beds in the new maternity ward has increased from fifteen to thirty-three. The ward now has four labour and delivery beds. It also includes a special care nursery with the capacity to hold eight babies and constructed to withstand category 4 and category 5 cyclones (Willie 2019).

On the occasion of the inauguration of this renovated maternity ward in November 2019, the Australian foreign minister, Marise Payne, declared that "there's not much more powerful demonstration of the partnership" between the two countries. She continued:

> Because health is a priority for both of us, Australia and for Vanuatu, to ensure that our communities are able to live healthy and productive lives, and over the last nine years, we have continued to invest in health service delivery and in public health administration and we have produced together in that partnership some really impressive results. (Willie 2019, n.p.)

She also highlighted some more positive outcomes of "the partnership": "No deaths from malaria since 2012. We have had 184 proud students graduated with diploma of nursing and 30 graduated with post diploma in midwifery" (Willie 2019, n.p.).

No doubt these improvements would please Dr. Freeman, who in the 1960s had advocated so strongly for improvements to the maternity ward in the mission hospital, as had his Presbyterian predecessors since 1937. Surely, they, too, would be outraged that it has taken this long to have a safe maternity ward in the area. The maternity ward, whose funding came from the PPB, among other sources, is also the progeny of, and reflects the reproduction of relations between, the tax haven set up by a colonial government that would not provide adequate infrastructure after the mission handed over medical services to it in the 1960s. I find it a sign of the times that after all of the labour of Ni-Vanuatu staff, missionaries and their supporters, colonial officials, and an overseas investment bank made a donation – not even all that sizable, given the profits it earns – and the ward was then deemed complete.

Vanuatu, while still occupying a place of isolation within a global imagination (e.g., Bjornum 2020; Gunia 2020) – a claim seemingly evidenced by the fact that the novel coronavirus was largely kept out of the archipelago in 2020 and 2021 – is not isolated from the effects of climate change, or the effects of the loss of tourism on its economy, or the effects of medical infrastructure being paid for by investment banks and Australian donations. It is estimated that over 60 per cent of Vanuatu's formal economy is dependent on tourism, and the border closures will have devastating effects. The borders opened again for the export of kava on 25 April 2020. Many Ni-Vanuatu working overseas were unable to return home for months.

By late May 2020, beautiful images of spectacularly large piles of yams began to circulate on social media. On 23 May, the *Vanuatu Daily Post* (2020a) reported the following:

> A total of 27 tonnes of soft yams from the Shepherd Islands of Buninga and Tongariki will be supplied to affected communities in the Northern provinces through the National Disaster Management Office (NDMO) and Ministry of Agriculture, Livestock, Forestry, Fisheries and Biosecurity (MALFFB)'s Food Security and Agriculture Cluster (FSAC).
>
> This is the first time ever in history for the Government, through NDMO and negotiations done by MALFFB FSAC team and the Vila community of Buninga and Tongariki, to purchase a huge quantity of yams at over VT5 million.
>
> The total number of yams supplied by these farmers was 9,165 yams sold at VT200 per kilo.

The Vanuatu Department of Laefstok (Livestock) later posted images of yams on its public Facebook group, along with the following caption:

> This "jaw-dropping" footage shows piles and piles of yams the tiny island of Tongariki supplied FSAC to redistribute to the affected population on Pentecost and Santo.
>
> The fact of the matter is no matter how small in size, perseverance drives these people to grow as many yams over the years for the sake of their own food security.
>
> Not knowing that one day, they will become the largest supplier of local fresh root crops to disaster zones in SANMA and PENAMA province.
>
> Seen here 22,000 kg of yams, will take two days and 70 villagers to transfer from beach to ship, destined for Pentecost then Santo and Malo over the weekend. (Vanuatu Department of Laefstok 2020, n.p.)

Reactions to this news on various public fora were enthusiastic. For example, on the same Department of *Laefstok* Facebook post, people left many laudatory comments, of which the following is but a sample:

> *Laef istp lo graon. Tuff!* [Life is in the land. Great!]

> Wow. I am humbled by this display of nationhood. Caring for each other [followed by a heart emoji].

> Thank God for the disaster faced because out of this *Hem i soem lo Yumi blo rimaidem yumi bageken se mani i stap lo mama graon mo taim umi wok tugeta long wan ting bae Umi save ajivim ol bigfala samting blo benefit blo everiwan.* [… He shows us to remind us again that money is in Mother Earth and when we work together on one thing we can achieve great things that benefit everyone].

> Well done *ol Families lo Tongariki from fruit blo hadwok blo everiwan we bae helpem ol Families blo Umi lo North* [followed by a folded hands emoji] [Well done Tongariki families, the fruit of everyone's handiwork will help the families of those of us in the North].

> *LONG GOD YUMI STANAP* [In God we stand up – this is the national motto of Vanuatu].

> Together we are strong [followed by ten Vanuatu flags].

Tannese people in the town of Luganville, on the northern island of Espiritu Santo, are also supporting the post–Cyclone Harold rebuilding effort. In the absence of shelter materials arriving from overseas to the areas levelled by the hurricane, the Santo Sunset Environment Network has established a "grassroots coconut weaving programme." Using traditional knowledge from Tanna, dried coconut fronds, not normally used on West Santo, are being repurposed for use as a possible shelter material (Fuatai 2020).

During these days and months from February to June 2020, I consumed as much social media coverage as I could. Despite the unprecedented nature of the disasters that threatened to alter reproduction, in the social media posts I have reproduced here, I see contemporary iterations of relations of reproduction that I have written about in this book.

Foreign governments, and now banks, appear as well-intentioned "partners" in the construction of the maternal health infrastructure begun by missionaries, colonial officials, and the Vanuatu government, and implemented by Ni-Vanuatu labour and care networks. Where once there were squabbles between French and British colonial authorities within the Condominium, there are now increasing disputes between Australia and China over who will deliver goods and expertise. As well, the familiar narrative of resilience is clearly apparent and considerably amplified by the overlapping impacts of the COVID-19 pandemic and climate change. The catch-all term "resilience" encompasses the networks and skills included among the "alternative indicators of well-being" or "subsistence" in the first census. Land continues to be the emblematic and material bedrock of survival.

The networks of land, knowledge, kin, and care that constitute the "resilience" of Pacific Islanders speak to the distributed and assisted processes of reproduction that I have tried to describe in this book. Reproduction takes place through infrastructure and environments as well as through bodies. That reproduction extends beyond the individual body has long been evident in Pacific contexts; as Jolly (2001, 2002) has argued, Ni-Vanuatu practices and values associated with reproduction involve both biological and social dimensions, as well as male and female work and presence. Reproduction takes place in bodies and also in relation to how the land is worked and made fertile. These processes are in turn connected to the reproduction of kinship ties that have assumed different inflections across the archipelago. The expansion of Christianity and colonialism meant that reproduction was more closely connected with women's work and identities (Jolly 1991). One political outcome of locating reproduction in bodies alone is that women are made responsible for reproduction, and population-level interventions are technological fixes for controlling biological reproduction or efforts to enhance wage labour in isolation from other aspects of social reproduction.

The episodes and moral figures I have described in this book have shown how reproduction occurs in a context in which people articulate and create futures in the face of existing and potential threats. Highlighting reproduction is a way to centre Ni-Vanuatu articulations of futures. I see future-facing work through the women who came to new courts to try and find a way to change divorce practices in the 1920s, and in the women who trained as nurses in the 1950s and 1960s and who cared for birthing women despite the challenging conditions in the hospital. I see Indigenous futurities through the Ni-Vanuatu who, in varying ways, insisted that land use and access were crucial issues in the 1960s. Futurities are also apparent in the young mothers who, after remaining quiet at home during their pregnancies, managed whatever relationships they needed for themselves and their children. Finally, I see Ni-Vanuatu futurities in the work of massage healers who taught me that well-being is connected to doing the right amount of work and especially the maintenance of reciprocal relations. Murphy's insights that open this chapter indicate that life exceeds "the infrastructures of its management and valuation" (2017, 141). In Vanuatu, indeed, despite a century of attempts to quantify and monetize the unpaid dimensions of reproduction in the name of governance, control, and improvement, the reproduction of life has continued to exceed economization, but not the need for reciprocal relations.

"How might we build new forms of life out of the old?" This question, as Murphy (2015a, 287) argues, must accompany the theorizing of reproduction. Murphy conceives of reproduction as a collection of relations that extend far beyond the common conceptualizations of body, sexual reproduction, birth, and childcare, and that cannot be subsumed by labour. She argues that reproduction must also encompass the unequal relationships surrounding the care and pollution of life on the molecular and the planetary scales, and "practices of subsistence, of co-existing with humans and non-humans, of flourishing together in sustainable conditions, of reconciliation and of pleasure and desire," and much more. In this expansive construction, "the politicization of the relations of reproduction is crucial to the critique of capitalism and the project of imagining and struggling for other worlds" (300–1). Histories of reproduction in places like Vanuatu show the urgency of this project – indeed, they have been showing us this since the early twentieth century, as anyone who is willing to see can tell you. The COVID-19 pandemic, coupled with the catastrophic effects of climate change, clearly brings this urgency into view, yet again.

The "partnerships" that produce infrastructure reflect a reproductive metaphor and worlding. The question that arises is what new partnerships and metaphors will emerge in the Anthropocene? In a gesture of solidarity, the government of Vanuatu sent 20 million vatu (CAD$250,000) to support Australia's efforts to fight their devastating bush fires in January 2020 (Ewart and Handly 2020; McArthy 2020). During March and April 2020, when many kinds of paid work were not possible, there were calls circulating on social media, clearly intended for audiences in affluent contexts in the Global North, to consider carefully whether we need to return to the pre-pandemic normal (e.g., Gambuto 2020). Can the future put us in better relations with non-humans? Is this the death knell of neoliberal capitalism? In the first months of the COVID-19 pandemic, critical calls to be attentive to the presence of "disaster capitalism" – whereby private corporations swoop down on weakened and vulnerable people and social systems to set up for-profit systems – fill my own and others' social media feeds (e.g., Alexander and Staley 2020; Klein 2020), alongside commentaries on the climate-change impacts that caused the disaster of Cyclone Harold.

The act of putting the flourishing of certain lives on hold so that others can flourish lays bare the failure to recognize reciprocal relations at many scales. Such acts recalled the challenge of the early twentieth century of how to attend to reproduction during and after devastating infectious disease epidemics in Vanuatu, as was the case in Indigenous societies in many places. That depopulation concerns have given way to concerns about rapid population growth could also be seen as a resurgence. It is not necessarily fertility that needs to be problematized, but other aspects of non-reciprocal relationships around land use, urbanization, underfunded infrastructure, and consumption patterns outside Vanuatu that are causing climate change. As the first months of 2020 demonstrated, relations of reproduction extending into the future will have to contend with climate change and food production – in addition to infectious diseases – in acute ways in the Pacific Islands. Pacific Islanders' care networks and their skilled knowledge of the land are laudable. But they are emphatically not isolated, and relations between the Pacific and elsewhere will need to be produced that allow all to flourish. New moral figures will flourish from this fertile space.

Appendix One

Ni-Vanuatu Population Size, 1850–2020

Pre-1850: estimated between 500,000 (Spriggs 1997, 261) and 650,000 (Speiser [1923] 1996, 39)
1910: 64,555 (Speiser [1923] 1996, 33)
1967: 70,837 (McArthur and Yaxley 1968, 23)
(Plus the approx. 1,406 people on Santo and Tanna who refused to participate)
1979: 112,596 (National Planning and Statistics Office 1983)
1989: 142,944 (VNSO 1993)
1999: 186,678 (VNSO 2001)
2009: 234,023 (VNSO 2010)
2016 Mini Census: 272,459 (VNSO 2016)
2020: 300,019 (VNSO 2021b)

Appendix Two

Overview of Biomedical Health Services in Vanuatu in 1954

Extracts from "General Review of Medical Services in the Anglo-French Condominium of the New Hebrides, 1954," by NHBS Medical Officer A.R. Mills (1954).

Administrative Organization and Institutions

The hygiene and public health of the Condominium are under the control of the Public Health Service of the Condominium. This service consists of a Chief Condominium Medical Officer, Condominium Medical Officers and sanitary formations. There are French or British National personnel acting on behalf of the Condominium. The Condominium itself runs two sanitary formations. The duties of the service through its officers are:

General oversight of public health in the Group;
Supervision of quarantine and preventive measures against epidemic and communicable diseases;
The giving of free medical attention to all natives of the New Hebrides.

The British National Medical Service

The British National Medical Service maintains one British Medical Officer, who is medical adviser to the British Residency and coordinator of British Services throughout the Group. The British Medical Officer undertakes rural health campaigns, Anti-Yaws, B.C.G. and health surveys throughout the group, and directs the activities of five Assistant Medical Practitioners in the British Service. The five Assistant Medical Practitioners maintain bush clinics in different parts of the group and do hygiene tours in their districts.

The British Authorities subsidise the medical services of the Presbyterian and Melanesian Missions. The Presbyterian Mission runs the Paton Memorial Hospital at Vila and a hospital at Lenakel, Tanna, both of which have resident European doctors. About thirty dressers are maintained in villages by the Mission and the training of nurses and dressers is carried out in Vila. The Melanesian Mission has a hospital and small leprosarium on Aoba run by New Zealand Nursing Sisters and an Assistant Medical Practitioner.

The Mission has also dressers stationed on other islands. Training of nurses and dressers is an important part of the work of the Melanesian Mission.

The Churches of Christ Mission has a small hospital with a European Nursing Sister in charge on Aoba.

The French National Medical Service

This is under the direction of a Major of the French Colonial Medical Service, who is also Chief Condominium Medical Officer.

There are four French National Hospitals, one at Vila, one at Santo and two on Malekula at Lamap and Norsup. The first two are well equipped and take European patients. The hospitals are staffed by three military doctors, one of whom is a surgeon, and by fourteen Sisters of the Society of Mary.

Condominium Medical Service

There are two hygiene squads (one each at Vila and Santo) which do anti-mosquito control work and general hygiene in the two urban centres, and in villages.

The Condominium has three Assistant Medical Practitioners on staff, two of which work in bush hospitals, the third assisting the French surgeon. The Condominium also maintains native dressers and dressing stations.

The Condominium subsidises through the Indigent Native Subsidy, British, French and Mission services for the treatment of New Hebridean natives.

Dentistry

There is one French private practitioner in Vila, and one native Assistant Dental Practitioner working at the Presbyterian Mission Hospital in Vila.

Sanitation

Vila and Santo are the only towns in the New Hebrides. Sanitation of these urban areas is governed by Joint Regulation and each is under the supervision of a Sanitary Inspector. Mosquitoes control in these: centres is of a reasonable standard. Native sanitation is governed by a special Joint Regulation, and the standard of native hygiene is progressively improving especially in the neighbourhood of the towns and rural clinics.

Medical and Auxiliary Personnel in the New Hebrides, 31st Dec., 1954

French Government Staff
Registered Medical Practitioners 4
Colonial midwife 1
Nursing Sisters (Society of Mary) 14

French Missions
Untrained Nurses 18

Condominium Government
Assistant Medical Practitioners 3
Dressers 11
Sanitary Inspectors 2
Hygiene Assistant 1

British Government
Registered Medical Practitioners 1
Assistant Medical Practitioners 5

British Missions
Registered Medical Practitioners 3
Midwives (trained nurses) 16
Untrained Nurses 49
Assistant Medical Practitioners 2
Assistant Dental Practitioner 1
Dressers 84
Private Dentist 1

Overall Total
Registered Medical Practitioners 8
Private Dentist 1

Assistant Medical Practitioners 10
Assistant Dental Practitioner 1
Trained Nurses 31
Untrained Nurses 67
Dressers 95
Sanitary Inspectors 2
Mental Hospital Guardian 1
Hygiene Assistants 1

Summary of British Medical Service & British Subsidised British Medical Service

British Medical Service
Personnel
a. One British Medical Officer.
b. Five Assistant Medical Practitioners.

Institutions
One bush hospital, Abwatuntora, Pentecost. One bush clinic, Saranabuga, Aoba.

British Subsidised Medical Service
Personnel
a. Three Medical Missionaries.
b. Six qualified nurses.
c. Two Assistant Medical Practitioners.

Institutions
a. Paton Memorial Hospital, Vila, Efate.
b. Presbyterian Hospital, Lenakel, Tanna.
c. Godden Memorial Hospital, Lolowai, Aoba. (Mel. Mission)
d. S.W. Bay Clinic.
e. Tongoa Clinic.

Rural Dispensaries
About 30 rural dispensaries are run by British Missions throughout the group, on the following islands:

Maewo	Mau
Vanua Lava	Lelepa
Nguna	Ambrym
Melekula	Espiritu Santo

Tongoa	Malo
Tongoa	Epi
Paama	Emai
Maskelynes	
Erromango	Tanna
Aneityum	Achin

Description of Medical Institutions

1. **Paton Memorial Hospital**, Vila, Efate. Presbyterian Mission
   Personnel
   (a) One medical missionary.
   (b) Three qualified midwives.
   (c) Six native nurses.
   (d) One native midwife.
   (e) One Assistant Medical Practitioner.
   (f) One Assistant Dental Practitioner.
   The hospital is in the process of rebuilding. It contains one ward for Europeans of five beds, and two wards for natives containing 19 beds. There is an operating theatre, X ray equipment and clinical laboratory. The hospital is equipped for all medical and surgical cases.

2. **Presbyterian Hospital, Lenakel Tanna.**
   Personnel
   (a) One medical missionary.
   (b) One qualified midwife.
   (c) Eight native nurses.
   The hospital contains one ward for Europeans containing two beds and two wards for natives with 18 beds. The hospital is equipped with an operating theatre. In addition there are native built houses around the hospital in which out-patients under treatment who come from a distance may stay.

3. **Godden Memorial Hospital, Lolowai, Aoba, Melanesian Mission.**
   Personnel
   (a) Two qualified midwives.
   (b) One Assistant Medical Practitioner.
   The hospital has one ward for Europeans of two beds and three wards for natives containing 15 beds. There is also a small leprosarium in two sections, one for nine male patients, and the other

for nine female patients. The lepers are housed in native huts in the hospital grounds. They have small beaches with canoes for fishing and gardens to cultivate.

4. The Clinic at Abwatuntora, Pentecost.

This is run by a British Service Assistant Medical Practitioner. There are two buildings of native construction, one a dispensary and the other a four roomed building for four male and four female beds with two maternity beds and labour ward.

5. The Clinic at Saranabuga, Aoba.

This is run by a British Service Assistant Medical Practitioner. It is a native structure, divided into two portions, one for consultations and one for dressings.

Notes

Preface

1 Ni-Vanuatu are the Indigenous citizens of Vanuatu.

Introduction

1 John Baker, William Rivers, and Felix Speiser, along with other medical researchers such as Sylvester Lambert and Patrick Buxton, offered recommendations for solving depopulation (Widmer 2012; Coghe and Widmer 2015). Notable among these was the provision of better medical care by the colonial authorities so as to reduce mortality rates. During the first decades of the twentieth century, the medical system in the New Hebrides was run by English-speaking Presbyterian and Anglican missionaries, whereas the French colonial authorities operated hospitals and small clinics. A Rockefeller Foundation–funded campaign to address hookworm (not a life-threatening disease) was one of the medical initiatives introduced in 1926. As well, Pacific Islanders trained at the Fiji School of Medicine and began coming to the New Hebrides in the 1926, though the first Ni-Vanuatu assistant physician only began working there in 1938.

2 To use the name Vanuatu to refer to events in the colonial period is technically an anachronism. It is, however, common in writings describing this period. In making this historiographical choice, I follow the conventions of various respected scholars (e.g., Jolly 1998; Meyerhoff 2002). Older people in Vanuatu would call the colonial period *taem blong tufala gavman* (*tufala gavman* refers to the Condominium, while *tufala* means "two") or *bifo independens* (before independence). When I discuss colonial and missionary activities (which formed the New Hebrides as a polity), I use the term "New Hebrides."

3 The New Hebrides Advisory Council, as will be discussed in chapter 3, was an unelected body that provided advice to the British and French. The

colonial officials were not obliged to take the recommendations, but often did. In 1957, the council was made up of "the two resident commissioners, two official members and 12 nominated representatives of the major ethnic groups: four British, four French and four ni-Vanuatu" (Gardner 2013, 127). In recognition of their role in the provision of health and education services, the major churches were part of the Advisory Council: the Anglican and Catholic Churches were represented by Bishop (later Archbishop) Rawcliffe and the Roman Catholic vicar general, respectively. By 1968, there were "14 indirectly elected private members: three British, three French and eight Melanesian" (Gardner 2013, 127–8).

4 The first long-term European settlement was founded by Nova Scotian Presbyterian missionaries (from what is now the Canadian province of Nova Scotia) on the southern island of Aneityum in 1848, followed by Efate in 1861. Anglicans began missions in the northern islands in the 1850s.

5 Because of the lack of data before 1967, it is difficult to adequately explain all the factors behind the growth in population. With regard to the fertility side of population growth, it is difficult to know whether the low birth rates in the early twentieth century were due to infant mortality, infanticide, or traditional means of fertility control.

6 Of course, the interaction between the public and science is important for understanding both the need for quantification in business and governance as well as in science (Porter [1995] 2020).

7 I follow a similar methodology to that employed by Margaret Rodman (2001) in her documentation of the colonial history of Vanuatu through a study of British colonial space.

8 The NHBS had provided some funds since the PMH officially opened; 150 pounds per year in 1911 (Miller 1986, 117). The NHBS gradually took over the hospital administration and finance in the 1960s.

1. "The Shortage of Women Is the Cause of These Courts": Imbalanced Sex Ratios, Native Courts, and Marriage Disputes Made Public, 1910–1950

1 The bride price remains a common practice in contemporary Vanuatu and is now set by the Vanuatu government at a sum of 80,000 vatu (about CAD$880). This chapter shows a colonial precedent for such state policies. The bride price continues to be a public matter whereby kinship is officially entangled with economics.

2 Plantation labour in the New Hebrides was organized in such a manner that Ni-Vanuatu men tended to work on plantations away from their villages in other parts of the archipelago during this period (Bedford 1973).

3 Copra is dried coconut, one of the few commodities at the time processed in Vanuatu for export.

4 Forsyth (2009, 73) writes that forms of leadership during the time of the Condominium varied. In "some places the traditional leaders continued to be responsible for law and order, while in others those who followed newer roads to authority found themselves in charge."

5 Harrison would go on to write *Savage Civilization* (1937) about his time in Vanuatu. The "savage civilization" to which he refers is the capitalist and colonial order expanding across the region.

6 Australian pounds were the official currency of Australia from 1910 to 1966.

7 Forsyth (2009) writes that the major changes to leadership during the Condominium era included the establishment of chiefly institutions, new opportunities for leadership within the Condominium and Christian institutions, and the reduction or elimination of Indigenous leadership paths, termed *nimangki*. The "native courts" that I describe here were part of the changes made to dispute-resolution mechanisms under the Condominium that Forsyth identifies, in particular the public meetings that preceded the courts. Ni-Vanuatu styles of leadership that emphasized oratory for winning consent (rather than using coercion) were the bedrock of these meetings.

8 In the New Hebrides, there were three legal codes in operation: French, British, and Native. Stevens (2017, 596) writes that "Although provisions for the creation of a Native code were established earlier, relevant legislation was not enacted until 1927–28 with the Joint Regulation No. 6 of 1927 (Code of Native Law), Joint Regulation No. 1 of 1928 (Native Criminal Code), and Joint Regulation No. 2 of 1928 (Institution of Native Courts)."

9 At this time, as schools were run by missionaries, Christian Ni-Vanuatu were likely to have some basic primary schooling.

10 The Joint Court was the forum for hearing cases concerning offences between settlers and those involving settlers and Ni-Vanuatu.

11 When a man wanted to end his marriage and took up temporary or more permanent residence with another woman, that did not trigger the same paper trail relating to the matter of the return of the bride price. This situation was recorded in the colonial files in a minute from a 1942 meeting of the Presbyterian Synod, whose members expressed "concern at the number of native women deserted by their husbands and left without support." The reply to this minute, which reflected a common bureaucratic rejoinder, was that, while more information was needed, at the moment nothing could be done, as the Native Code did not have provisions for such circumstances (Paton 1942). Women's desire to leave became particularly visible because it was the *bride price* (not a dowry, for example) that needed to be returned. Because the content of the files dealt with contentious elements, the marriages recorded in the files mainly dealt with cases where the woman wanted to end the marriage.

2. "The Nurses Looked Out for Us!": Hospital Births, Relational Infrastructures, and Public Concerns, 1950–1970

1 I use pseudonyms throughout this chapter whenever quoting from interviews and conversations with my interlocutors.
2 At the beginning of my research, I asked whether I could record these interviews. The first few women were uncomfortable with this, so I just stopped using a recording device and took notes during the interviews instead. All of the interviews were in Bislama, but I took notes primarily in English and thus the content recounted here is based on my own translations.
3 There is a long-standing connection between "population" and "economy" in European modernity. It goes back at least to the eighteenth-century French physiocrats and their concern for the wealth of the population and agricultural productivity, Malthus's writings on resources and population, and Marx's conception of surplus population in respect to labour.
4 Birth control was also another; this is discussed in chapter 4.
5 Pandanus is a tree or shrub, the fronds of which are harvested and dried by Ni-Vanuatu and subsequently woven into durable mats that serve many household and ceremonial purposes.
6 The first Christian church in Vanuatu was permanently established (after several years of attempts) on the island of Aneityum in 1848, followed by one on Efate in 1861 (Miller 1978). For a detailed history of the medical work of the Presbyterian Church in Vanuatu, see Miller (1986).
7 The launch did not leave the wharf by the Port Vila market, where one catches a boat to Iririki island from the Port Vila waterfront today.
8 Pyatt is fondly remembered by many in both New Zealand and Vanuatu. An apartment building in the retirement complex where she lived in the last years of her life is named in her honour (Selwyn Foundation 2017).
9 Those present were Sisters Heard, Pyatt, Atteneave, and Edgar, and Drs. Coulter, Mackereth, Freeman, de Wilde, Pennington, LeClerq, Loisson, Ramm, Greenough, and Joeli.
10 This includes the large villages surrounding Port Vila: Mele, Maat, Fila, Pango, Erakor, and Eratap (McArthur and Yaxley 1968, 30).
11 Citing Foster, Rawlings (1999b, 42) writes that between 1972 and 1974 the expatriate population of Port Vila tripled and thirteen overseas banks opened.
12 Following the completion of her nursing training, as was the case with a handful of promising students, Madeleine was sent for six months of specialized training in Sydney, Australia.
13 With Drs. Freeman, Jameson, and Nell Cruickshank.

3. "It Will Help Planning for the Future": Making Men's and Women's "Subsistence" Public Knowledge in the First Census, 1966–1967

1 Words by Mrs. W. Camden of south Santo, sung to the tune of "Black Jack from Gundagai" (McArthur and Yaxley 1968, ii). I have reproduced this source as it was written. Bislama spelling has changed considerably since 1968. The song now seems to be rendered as "Flash Jack from Gundagai."

2 Australia's National Security Intelligence Agency (comparable to the US Department of Homeland Security and CIA or the British MI5) held a classified file on McArthur. The file contained her application to work in the United Nations Technical Assistance program in Indonesia as well as a request for permission to pass through Papua on her way to the Solomon Islands in the 1950s.

3 Though I intend to mark this word as a historically and socially specific term for the entire chapter, I only use quotation marks in this opening section.

4 Now called the Pacific Community. During this time, the SPC also published a series of technical reports on nutrition, food production, and agriculture.

5 See Native Census Meeting Minutes (1963) in the list of unpublished references.

6 This mine operated from 1962 to 1979.

7 When the census was complete, McArthur even went so far as to claim that the drop in population had been greatly overstated, writing "there can be little doubt that the extent of the 'depopulation' in the sense of an absolute net loss of population has been exaggerated (perhaps even grossly) by biases and misinterpretations" (McArthur and Yaxley 1968, viii). With the accumulation of archaeological evidence pertaining to population size in Vanuatu's deep history, McArthur's scepticism needs better consideration (e.g., Flexner and Spriggs 2017). For example, Spriggs (2007) estimates there was a pre-contact population of 6,000–8,000 on Aneityum, while McArthur and Yaxley (1968, 12) estimated 3,500 (this population went as low as 200 in the 1930s).

8 For more details on how race and citizenship were enacted in colonial Vanuatu, see Rawlings (2019), Rodman (2003), Stevens (2017), and Widmer (2014).

9 The Joint Court was established in the early days of the Condominium, mainly to deal with the large number of land disputes and to serve as a colonial mechanism whereby non-Indigenous individuals and entities could register legal title to land (Rodman 2001, 29).

10 The SFNH was initiated when the French government took over the failing Compagnie Caledonienne des Nouvelles-Hébrides from John Higginson, an Irishman who had become a French citizen. Higginson had

acquired a great deal of land from struggling British settlers, as well as from islanders. By 1894, the SFNH owned more than 55 per cent of the cultivated land in the islands (Sénat Français, n.d.).

11 For an extended description and analysis, see Beasant (1984).

4. "I Just Wanted to Be Invisible": "Young Mothers" from Global Discourse to Village Experience, 2010–2020

1 I indicate young mothers with quotation marks here, but not throughout the chapter. The phrase should always be taken in context as an emic term.

2 *Kastom* refers to knowledge and practices that differentiate Ni-Vanuatu from outsiders. *Kastom* also varies within Vanuatu. There is a large anthropological literature on *kastom* that I discuss at greater length in chapter 5.

3 Hugo's solution to the population pressure is to promote Pacific Islanders' temporary migration to Australia, as New Zealand has done through its Recognized Seasonal Employment scheme, which could not only prove a benefit to Australia, but also would be a viable part of the development plan of the sending nation (Razak and Hugo 2012; Hugo 2009).

4 *State of Pacific Youth Report* (2019) is the result of a collaboration between several prominent international organizations active in the region: the UN Pacific Regional Youth Theme Group (namely, ILO, UNDP, UNFPA, with UNESCO as chair and UNICEF, UN Women, and WHO as co-chairs), the Pacific Community, the Pacific Youth Council, and the Commonwealth Secretariat.

5 The eighteen to twenty-four age range is used because it is a report about Pacific youth.

6 The remainder of the goals are described as follows: "2. Reduce infant, child and maternal mortality and morbidity; 3. Manage rural-urban migration and urbanization; 4. Improve the availability of data and the integration of population into sector plans and national development strategies; 5. Promote gender equality and reduce gender based violence; 6. Reduce unemployment and underemployment rates among youth and legislate working age (18–35 years); 7. Reduce hardship and poverty among the elderly, widowers, people with disability, and other vulnerable people" (Vanuatu Department of Strategic Policy, Planning and Aid Coordination 2011, 43).

7 In 2000, 189 countries came together and made a promise to free the people of the world from extreme poverty and deprivation by 2015. This commitment was formalized in the publication of the United Nations Millennium Declaration, which commited world leaders to combating poverty, hunger, disease, illiteracy, discrimination against women, and environmental degradation.

In order to achieve these aims, eight Millennium Development Goals (MDGs) were put in place as a framework for actions to be taken and against which progress would be measured:

1 Eradicate extreme poverty and hunger
2 Achieve universal primary education
3 Promote gender equality and empower women
4 Reduce child mortality
5 Improve maternal health
6 Combat HIV/AIDS, malaria and other diseases
7 Ensure environmental sustainability
8 Develop a global partnership for development

The MDGs were inter-dependent; all the MDGs influence health, and health influences all the MDGs. Better health enables children to learn and adults to earn. Gender equality is essential to the achievement of better health. Reducing poverty, hunger and environmental degradation positively influences, but also depends on, better health (Maternity Worldwide, n.d.).

8 *Strong hed* is frequently used to describe children who do not do as they are told or adults who do not follow expectations.
9 In writing about these interviews and the stories that my interlocutors shared with me, I have chosen to include only very general details in order to protect these women's privacy and to prevent further re-stigmatization after their circumstances have changed.

5. "Well-Being for Melanesia": Alternative Indicators, Massage Healers, and Knowledge of Reciprocal Relationships, 2010–2021

1 I use pseudonyms throughout this chapter whenever quoting from interviews and conversations with my interlocutors. Readers in Vanuatu can consult oral histories deposited at the Vanuatu Cultural Centre for real names, according to the wishes of the storytellers.
2 Such resorts, constructed with Australian and European capital, are held on seventy-five-year leases. As discussed elsewhere in this book, the *kastom* (traditional) owners are kin-based groups. Sometimes, land access is shaped by matrilineal relations who form a *naflak*, but increasingly by father-son relationships called *blad laen*. These owners have the right to reclaim ownership at the end of the lease. In the time leading up to my fieldwork in 2010, prominent members of families, often men, had made the decision to turn over large swathes of ancestral land. It has meant the posting of court decisions settling land disputes between local families on store windows in Pango. It has meant the documentation of genealogies on paper, at times laminated in special albums and held as precious

objects that record who belongs within the *blad laen*. It has meant frequent disputes over whether land title belonged to *kastom* owners, as defined by matrilineal clans or blood lines. And for me, it meant, methodologically, that oral histories, often told genealogically, were challenging to inquire about.

3 As mentioned at other points in this book, rural contexts on the island of Efate or other islands in Vanuatu would look very different in terms of access to wage labour, mass-produced goods, and proximity to medical institutions.

4 Residents of the Port Vila area may also have learned about the role of women in the traditional economy by visiting an exhibition focusing on this topic that was organized at the Vanuatu Cultural Centre.

5 In October 2015, Moana Carcasses was sentenced to four years in prison for corruption (Australian Broadcasting Corporation 2015).

6 I describe some of the coercive aspects of medical and non-medical care/concern for young women in other chapters.

7 This narrative was not recorded; it is reconstructed here from the detailed notes I took at the time.

Epilogue: Relations of Reproduction and Survival in the Anthropocene

1 For more on this program, see Gibson and Bailey (2021).

References

Abernethy, Virginia. 1995. "The Demographic Transition Model: A Ghost Story." *Population and Environment* 17, no. 1 (September): 3–6. https://doi.org/10.1007/BF02208273.

Abu-Lughod, Lila. 2013. *Do Muslim Women Need Saving?* London: Harvard University Press.

Adams, Vincanne. 2016a. "Introduction." In Adams 2016b, 1–18.

–, ed. 2016b. *Metrics: What Counts in Global Health*. Durham, NC: Duke University Press.

Adelson, Naomi. 2000. *"Being Alive Well": Health and the Politics of Cree Well-Being*. Toronto: University of Toronto Press.

Ahmed, Leila. 1993. *Women and Gender in Islam: Historical Roots of a Modern Debate*. New Haven, CT: Yale University Press.

Aldrich, Robert. 1990. *The French Presence in the South Pacific, 1842–1940*. London: Palgrave Macmillan.

– 1993. *France and the South Pacific since 1940*. Honolulu: University of Hawai'i Press.

Alexander, Chloe, and Anna Staley. 2020 "Disaster Capitalism: Coronavirus Crisis Brings Bailouts, Tax Breaks and Lax Environmental Rules to Oilsands." *The Conversation*, 29 April 2020. https://theconversation.com/disaster-capitalism-coronavirus-crisis-brings-bailouts-tax-breaks-and-lax-environmental-rules-to-oilsands-135996.

Anand, Nikhil. 2018. "A Public Matter: Water, Hydraulics, Biopolitics." In *The Promise of Infrastructure*, edited by Nikhil Anand, Akhil Gupta, and Hannah Appel, 155–74. Durham, NC: Duke University Press.

Andersen, Barbara. 2016. "Temporal Circuits and Social Triage in a Papua New Guinean Clinic." In "Temporality as a Lens for NGO Studies," edited by Veronica Davidov and Ingrid Nelson. Special issue, *Critique of Anthropology* 36, no. 1 (March): 13–26. https://doi.org/10.1177%2F0308275X15617306.

Anderson, Benedict. 2001. *Imagined Communities: Reflections on the Origin and Spread of Nationalism*. London: Verso.

Anderson, Warwick. 2009. "Modern Sentinel and Colonial Microcosm: At the Philippine General Hospital." In "Public Health in History." Special issue, *Philippine Studies* 57, no. 2 (June): 153–77. https://www.jstor.org/stable/42634007.

Appadurai, Arjun. 1993. "Number in the Colonial Imagination." In *Orientalism and the Postcolonial Predicament: Perspectives on South Asia*, edited by Carol A. Breckenridge and Peter van der Veer, 314-39. Philadelphia: University of Pennsylvania Press.

Argast, Regula, Corinna Unger, and Alexandra Widmer. 2016. "Twentieth Century Population Thinking: An Introduction." In Population Knowledge Network 2016, 1–10.

Asad, Talal. 1992. "Conscripts of Western Civilization?" In *Dialectical Anthropology: Essays in Honor of Stanley Diamond*, vol. 1, edited by Christine Gailey, 333–51. Gainesville: University Press of Florida.

– 1996. "Comments on Conversion." In *Conversion to Modernities: The Globalization of Christianity*, edited by Peter van der Veer, 263–75. New York: Routledge.

– 2002. "Ethnographic Representation, Statistics, and Modern Power." In *From the Margins: Historical Anthropology and Its Futures*, edited by Brian K. Axel, 66–94. Durham, NC: Duke University Press.

Australian Broadcasting Corporation. 2015. "Vanuatu Court Sentences MPs, Including Former PMs Carcasses and Vohor, to Jail for Corruption." *ABC News*, 23 October 2015. https://www.abc.net.au/news/2015-10-22/vanuatu-mps-including-moana-carcasses-and-serge-vohor-sentenced/6875566.

Axelson, Per, and Peter Sköld. 2011. *Indigenous Peoples and Demography: The Complex Relation between Identity and Statistics*. New York: Berghahn Books.

Baker, John R. 1928. "Depopulation in Espiritu Santo, New Hebrides [with Plate XXIV]." *Journal of the Royal Anthropological Institute of Great Britain and Ireland* 58 (January–June): 279–303. https://www.jstor.org/stable/4619534.

Bamford, Sandra. 2007. *Biology Unmoored: Melanesian Reflections on Life and Biotechnology*. Berkeley: University of California Press.

Barcelos, Chris. 2020. *Distributing Condoms and Hope: The Racialized Politics of Youth Sexual Health*. Berkeley: University of California Press.

Bashford, Alison. 2014. *Life on Earth: Geopolitics and the World Population Problem*. New York: Columbia University Press.

Battaglia, Debbora. 2017. "Aeroponic Gardens and Their Magic: Plants/Persons/Ethics in Suspension." *History and Anthropology* 28, no. 3 (May): 263–92. https://doi.org/10.1080/02757206.2017.1289935.

Bear, Laura. 2013. "'This Body Is Our Body': Vishwakarma Puja, the Social Debts of Kinship, and Theologies of Materiality in a Neoliberal Shipyard." In McKinnon and Cannell 2013, 166–89.

Bear, Laura, Karen Ho, Anna Tsing, and Sylvia Yanagisako. 2015. "Gens: A Feminist Manifesto for the Study of Capitalism." *Theorizing the Contemporary* (blog), Society for Cultural Anthropology, 30 March 2015. https://culanth.org/fieldsights/652-gens-a-feminist-manifesto-for-the-study-of-capitalism.

Beasant, John. 1984. *The Santo Rebellion: An Imperial Reckoning*. Honolulu: University of Hawai'i Press.

Bedford, Richard D. 1973. *New Hebridean Mobility: A Study of Circular Migration*. Canberra: Australian National University.

Bhatia, Rajani, Jade S. Sasser, Diana Ojeda, Anne Hendrixson, Sarojini Nadimpally, and Ellen E. Foley. 2020. "A Feminist Exploration of 'Populationism': Engaging Contemporary Forms of Population Control". *Gender, Place & Culture* 27, no. 3 (March): 333–50. https://doi.org/10.1080/0966369X.2018.1553859.

Bhattacharya, Tithi, ed. 2017. *Social Reproduction Theory: Remapping Class, Recentering Oppression*. London: Pluto Press. https://doi.org/10.2307/j.ctt1vz494j.

– 2020. "Social Reproduction and the Pandemic, with Tithi Bhattacharya." By Sarah Jaffe. *Dissent*, 2 April 2020. https://www.dissentmagazine.org/online_articles/social-reproduction-and-the-pandemic-with-tithi-bhattacharya.

Biehl, Joao. 2005. *Vita: Life in a Zone of Social Abandonment*. Berkeley: University of California Press.

Biruk, Crystal. 2012. "Seeing Like a Research Project: Producing 'High-Quality Data' in AIDS Research in Malawi." In "Enumeration, Identity, and Health," edited by Thurka Sangaramoorthy and Adia Benton. Special issue, *Medical Anthropology: Cross-Cultural Studies in Health and Illness* 31, no. 4 (July): 347–66. https://doi.org/10.1080/01459740.2011.631960.

– 2018. *Cooking Data: Culture and Politics in an African Research World*. Durham, NC: Duke University Press. https://doi.org/10.1215/9780822371823.

Bjornum, Yasmine. 2020. "'If It Comes, It Will Be a Disaster': Life in One of the Only Countries without Coronavirus." *The Guardian*, 7 April 2020. https://www.theguardian.com/world/2020/apr/08/if-it-comes-it-will-be-a-disaster-life-in-vanuatu-one-of-the-only-countries-without-coronavirus.

Bledsoe, Caroline H. 2002. *Contingent Lives: Fertility, Time, and Aging in West Africa*. Chicago: University of Chicago Press.

Blundo, Giorgio. 2014. "Seeing Like a State Agent: The Ethnography of Reform in Senegal's Forestry Services." In *States at Work in West Africa: Dynamics of African Bureaucracies*, edited by Thomas Bierschenk and Jean-Pierre Olivier de Sardan, 69–89. Boston: Brill.

Boddy, Janice. 2007. *Civilizing Women: British Crusades in Colonial Sudan*. Princeton, NJ: Princeton University Press.

Bolton, Lissant. 2003a. "Gender, Status and Introduced Clothing in Vanuatu." In *Clothing the Pacific*, edited by Chloë Colchester, 119–39. Oxford: Berg.

– 2003b. *Unfolding the Moon: Enacting Women's Kastom in Vanuatu*. Honolulu: University of Hawai'i Press.

Bonnemaison, Joël. 1976. "Circular Migration and Uncontrolled Migration in the New Hebrides: Proposals for an Effective Urban Migration Policy." *South Pacific Bulletin* 26, no. 4 (Fourth Quarter): 7–13.

Bourdy, Genevieve, and Annie Walter. 1992. "Maternity and Medicinal Plants in Vanuatu: The Cycle of Reproduction." *Journal of Ethnopharmacology* 37, no. 3 (October): 179–96. https://doi.org/10.1016/0378-8741(92)90033-n.

Bremner, Brian. 2017. "Final Days of a Tax Haven." *Bloomberg Markets*, 15 November 2017. https://www.bloomberg.com/news/features/2017-11-15/the-final-days-of-a-tax-haven.

Briggs, Laura. 2017. *How All Politics Became Reproductive Politics: From Welfare Reform to Foreclosure to Trump*. Berkeley: University of California Press.

Broome, André, and Leonard Seabrooke. 2012. "Seeing Like an International Organisation." In "Seeing Like an International Organisation," edited by André Broome and Leonard Seabrooke. Special issue, *New Political Economy* 17, no. 1 (February): 1–16. https://doi.org/10.1080/13563467.2011.569019.

Brunson, Jan. 2019. "Tool of Economic Development, Metric of Global Health: Promoting Planned Families and Economized Life in Nepal." In "Behind the Measures of Maternal and Reproductive Health: Ethnographic Accounts of Inventory and Intervention," edited by Jan Brunson and Siri Suh. Special issue, *Social Science & Medicine* 254 (June): 1–9, article 112298. https://doi.org/10.1016/j.socscimed.2019.05.003.

Brunson, Jan, and Siri Suh. 2020. "Behind the Measures of Maternal and Reproductive Health: Ethnographic Accounts of Inventory and Intervention." In "Behind the Measures of Maternal and Reproductive Health: Ethnographic Accounts of Inventory and Intervention," edited by Jan Brunson and Siri Suh. Special issue, *Social Science & Medicine* 254 (June): 1–6, article 112730. https://doi.org/10.1016/j.socscimed.2019.112730.

Butt, Leslie. 1998. "The Social and Political Life of Infants among the Baliem Valley Dani, Irian Jaya." PhD diss., McGill University. https://escholarship.mcgill.ca/concern/theses/5h73px75f.

Callaci, Emily. 2020. "On Acknowledgments." *American Historical Review* 125 (1): 126–31. https://doi.org/10.1093/ahr/rhz938.

Campbell, Ian. 2006. "More Celebrated than Read: The Work of Norma McArthur." In *Texts and Contexts: Reflections in Pacific Islands Historiography*, edited by Doug Munro and Brij V. Lal, 98-110. Honolulu: University of Hawai'i Press.

Cheer, Joseph M., Keir J. Reeves, and Jennifer H. Laing. 2013. "Tourism and Traditional Culture: Land Diving in Vanuatu." *Annals of Tourism Research* 43 (October): 435–55. https://doi.org/10.1016/j.annals.2013.06.005.

Coghe, Samuel. 2022. Population Politics in the Tropics: Demography, Health and Transimperialism in Colonial Angola. Cambridge: Cambridge University Press.

Coghe, Samuel, and Alexandra Widmer. 2015. "Colonial Demography: Discourses, Rationalities, Methods." In Population Knowledge Network 2016, 37–64.

Cohn, Bernard. 1987. *An Anthropologist among the Historians and Other Essays*. Oxford: Oxford University Press.

Collier, Jane, Michelle Z. Rosaldo, and Sylvia Yanagisako. 1992. "Is There a Family? New Anthropological Views." In Thorne and Yalom 1992, 31–48.

Cosminsky, Sheila. 2016. *Midwives and Mothers: The Medicalization of Childbirth on a Guatemalan Plantation*. Austin: University of Texas Press.

Connelly, Matthew J. 2008. *Fatal Misconception: The Struggle to Control World Population*. Cambridge, MA: Belknap Press of Harvard University Press.

Córdova, Isabel M. 2017. *Pushing in Silence: Modernizing Puerto Rico and the Medicalization of Childbirth*. Austin: University of Texas Press.

Cummings, Maggie. 2005. "Who Wears the Trousers in Vanuatu?" In *Autoethnographies: The Anthropology of Academic Practices*, edited by Anne Meneley and Donna J. Young, 51–64. Peterborough, ON: Broadview Press.

– 2008. "The Trouble with Trousers: Gossip, Kastom, and Sexual Culture in Vanuatu." In *Making Sense of AIDS: Culture, Sexuality and Power in Melanesia*, edited by Leslie Butt and Richard Eves, 133–49. Honolulu: University of Hawai'i Press.

Davis-Floyd, Robbie. 2004. *Birth as an American Rite of Passage*. 2nd ed. Berkeley: University of California Press.

– 2018. *Ways of Knowing About Birth: Mothers, Midwives, Medicine, and Birth Activism*. Long Grove, IL: Waveland Press.

Davis-Floyd, Robbie, with Elizabeth Davis. 2018. "Intuition as Authoritative Knowledge in Midwifery and Homebirth." In Davis-Floyd 2018, 189–220.

DeLisle, Christine Taitano. 2015. "A History of Chamorro Nurse-Midwives in Guam and a 'Placental Politics' for Indigenous Feminism." *Intersections: Gender and Sexuality in Asia and the Pacific* 37 (March). http://intersections.anu.edu.au/issue37/delisle.htm.

Diocese of Melanesia. 1943. "Mothercraft Training and Infant Welfare." Ch. 10, In the Solomons and Other Islands of Melanesia. Sydney: Langlea Printery. http://anglicanhistory.org/oceania/solomons1943/.

Dörnemann, Maria, and Teresa Huhle. 2016. "Population Problems in Modernization and Development: Positions and Practices." In Population Knowledge Network 2016, 142-72.

Duclos, Vincent, and Tomás Sánchez Criado. 2020. "Care in Trouble: Ecologies of Support from Below and Beyond". *Medical Anthropology Quarterly* 34, no. 2 (June): 153–73. https://doi.org/10.1111/maq.12540.

Erikson, Susan L. 2016. "Metrics and Market Logics of Global Health." In Adams 2016b, 147–62.

Eves, Richard. 2019. "'Full Price, Full Body': Norms, Brideprice and Intimate Partner Violence in Highlands Papua New Guinea." *Culture, Health & Sexuality* 21, no. 12 (December): 1367–80. https://doi.org/10.1080/13691058.2018.1564937.

Ewart, Richard, and Erin Handley. 2020. "Pacific Nations Vanuatu and PNG Pledge Aid for Australia's Bushfires." *ABC News*, 6 January 2020. https://www.abc.net.au/news/2020-01-06/pacific-nations-pledge-aid-for-australias-bushfires/11844008.

Ferdinand, Ursula, and Petra Overath. 2016. "Organizations and Networks of Population Thinking in the First Half of the Twentieth Century." In Population Knowledge Network 2016, 65–89.

Finau, Silia Pa'usisi, Mele Katea Paea, and Martyn Reynolds. 2022. "Pacific People Navigating the Sacred Vā to Frame Relational Care: A Conversation between Friends across Space and Time." *Contemporary Pacific: A Journal of Island Affairs* 34 (1): 135–65. https://doi.org/10.1353/cp.2022.0006.

Flexner, James, and Matthew Spriggs. 2017. "When Early Modern Colonialism Comes Late: Historical Archaeology in Vanuatu." In *Historical and Archaeological Perspectives on Early Modern Colonialism in Asia-Pacific: The Southwest Pacific and Oceanian Regions*, edited by María Cruz Berrocal and Cheng-Hwa Tsang, 57–91. Gainesville: University Press of Florida.

Folbre, Nancy. 2002. *The Invisible Heart: Economics and Family Values*. 1st ed. New York: New Press.

– 2008. *Valuing Children: Rethinking the Economics of the Family*. Cambridge, MA: Harvard University Press.

Forsyth, Miranda. 2009. *A Bird that Flies with Two Wings: Kastom and the State Justice Systems in Vanuatu*. Canberra: ANU Epress.

Foucault, Michel. 1997. *Ethics: Subjectivity and Truth*. Edited by Paul Rabinow. Translated by Robert Hurley. New York: New Press.

Freeman, Edward Alan. 2006. *Doctor in Vanuatu*. Suva: IPS Publications, University of the South Pacific.

Fuatai, Teuila. 2020. "Coconut Resurgence in Cyclone-Devastated Vanuatu." *The Newsroom*, 25 May 2020. https://www.newsroom.co.nz/2020/05/25/1197711/coconut-resurgence-in-cyclone-devastated-vanuatu.

Gambuto, Julio V. 2020. "Prepare for the Ultimate Gaslighting: You Are Not Crazy, My Friends." *Forge*, 10 April 2020. https://forge.medium.com/prepare-for-the-ultimate-gaslighting-6a8ce3f0a0e0.

Gardner, Helen. 2013. "Praying for Independence: The Presbyterian Church in the Decolonisation of Vanuatu." In "Decolonisation in Melanesia: Global, National and Local Histories," edited by Helen Gardner and Christopher Waters. Special issue, *Journal of Pacific History* 48, no. 2 (January): 122–43. https://doi.org/10.1080/00223344.2013.781761.

Geismar, Haidy. 2013. *Treasured Possessions: Indigenous Interventions into Cultural and Intellectual Property*. Durham, NC: Duke University Press.

Gibson, John, and Rochelle-Lee Bailey. 2021. "Seasonal Labor Mobility in the Pacific: Past Impacts, Future Prospects." Asian Development Review 38, no. 1 (March): 1–31. https://doi.org/10.1162/adev_a_00156.

Gitelman, Lisa. 2013. Raw Data" Is an Oxy-Moron. Cambridge, MA: MIT Press. https://doi.org/10.7551/mitpress/9302.001.0001.

Government of Vanuatu. 2017. *Vanuatu 2030: The People's Plan. National Sustainable Development Plan 2016 to 2030, Monitoring and Evaluation Framework*. Port Vila: Department of Strategic Policy, Planning and Aid Coordination. https://www.gov.vu/images/publications/Vanuatu2030-EN-FINAL-sf.pdf.

Grace, Robert F. 2002. "Population, Public Health and Tubal Ligation in Vanuatu." In "Emergency Health in the Pacific," edited by Mark Keim and Sitaleki A. Finau. Special issue, *Pacific Health Dialog* 9, no. 1 (March): 17–20.

Greenhalgh, Susan, ed. 1995. *Situating Fertility: Anthropology and Demographic Inquiry*. Cambridge: Cambridge University Press.

– 1996. "The Social Construction of Population Science: An Intellectual, Institutional, and Political History of 20th Century Demography." *Comparative Studies in Society and History* 38, no. 1 (January): 26-66. https://doi.org/10.1017/S0010417500020119.

– 2008. *Just One Child Science and Policy in Deng's China*. Berkeley: University of California Press.

Gregory, Christopher. (1982) 2015. *Gifts and Commodities*. 2nd ed. London: HAU Books.

Guha, Ambalika. 2018. *Colonial Modernities: Midwifery in Bengal, 1860–1947*. London: Routledge.

Gunia, Amy. 2020. "This Tiny Nation Has Zero Coronavirus Cases. After a Devastating Cyclone, It's Refusing Foreign Aid Workers to Keep It That Way." *Time*, 17 April 2020. https://time.com/5820382/coronavirus-cyclone-harold-vanuatu/.

Haberkorn, Gerald. 2007–8. "Pacific Islands' Population and Development: Facts, Fictions and Follies." *New Zealand Population Review* 33–34:95–127.

Hacking, Ian. 1993. *The Taming of Chance*. Cambridge: Cambridge University Press.

– 2006. "Making Up People." *London Review of Books* 28, no. 16 (7 August 2006): 23–6.

Harradine, Mark. 2014. "Alienating Customary Land: People of the Land and People of Property in Vanuatu." PhD diss., Australian National University.

Harrison, Tom. 1937. *Savage Civilization*. London: Gollancz.

Hartmann, Betsy. 1995. *Reproductive Rights and Wrongs: The Global Politics of Population Control*. Boston: South End Press.

Hartmann, Heinrich, and Corinna R. Unger. 2016. "Family, Sexuality, and Gender." In Population Knowledge Network 2016, 115–42.

Hattori, Anne Perez. 2006. "'The Cry of the Little People of Guam': American Colonialism, Medical Philanthropy, and the Susana Hospital for Chamorro Women 1898–1941." In "History, Health, and Hybridity," edited by Judith A. Bennett, Barbara Brookes, and Annie Stuart. Special issue, *Health and History* 8 (1): 4–26. https://doi.org/10.2307/40111527.

Hawka, Charles. 2019. "50 Years Since Laying of Vila Central Hospital Foundation." *Vanuatu Daily Post*, 14 September 2019. https://dailypost.vu/news/50-years-since-laying-of-vila-central-hospital-foundation/article_479080b3-189d-5603-aaf8-3cfc2f799778.html.

Hendrixson, Anne, and Diana Ojeda. 2020. "Population." *Uneven Earth: Where the Ecological Meets the Political*, 20 July 2020. http://unevenearth.org/2020/07/population/.

Henley, Megan M. 2015. "Alternative and Authoritative Knowledge: The Role of Certification for Defining Expertise among Doulas." *Social Currents* 2, no. 3 (September): 260–79. https://doi.org/10.1177%2F2329496515589851.

Hess, Sabine. 2009. *Person and Place: Ideas, Ideals and Practice of Sociality on Vanua Lava, Vanuatu*. New York: Berghahn Books.

Hochschild, Arlie. 2012. *The Managed Heart: Commercialization of Human Feeling*. Berkeley: University of California Press.

Holmes, Teresa. J. 2009. "When Blood Matters: Making Kinship in Colonial Kenya." In *Kinship and Beyond: The Genealogical Model Reconsidered*, edited by Sandra C. Bamford and James Leach, 50–83. New York: Berghahn Books.

Hoodfar, Homa, Sheila McDonough, and Sajida Alvi, eds. 2003. *The Muslim Veil in North America: Issues and Debate*. Toronto: Women's Press.

Hugo, Graeme. 2009. "Best Practices in Temporary Labour Migration for Development: A Perspective from Asia and the Pacific." *International Migration* 47, no. 5 (December): 23–74. https://doi.org/10.1111/j.1468-2435.2009.00576.x.

Hull, Elizabeth. 2012. "Paperwork and the Contradictions of Accountability in a South African Hospital." *Journal of the Royal Anthropological Institute*

18, no. 3 (September): 613–32. https://doi.org/10.1111/j.1467-9655.2012.01779.x.
Hull, Matthew. 2012a. "Documents and Bureaucracy." *Annual Review of Anthropology* 41 (October): 251–67. https://doi.org/10.1146/annurev.anthro.012809.104953.
– 2012b. *Government of Paper: The Materiality of Bureaucracy in Urban Pakistan*. Berkeley: University of California Press.
Hunt, Nancy Rose. 1997. "'Le bébé en brousse': European Women, African Birth Spacing, and Colonial Intervention in Breast Feeding in the Belgian Congo." In *Tensions of Empire: Colonial Cultures in a Bourgeois World*, edited by Frederick Cooper and Ann Laura Stoler, 287–321. Oakland: University of California Press.
– 1999. *A Colonial Lexicon: Of Birth Ritual, Medicalization, and Mobility in the Congo*. Durham, NC: Duke University Press.
Jackson, Albert L. 1972. "Towards Political Awareness in the New Hebrides." *Journal of Pacific History* 7 (1): 155-62. https://doi.org/10.1080/00223347208572207.
Johnson-Hanks, Jennifer. 2008. "Demographic Transitions and Modernity." *Annual Review of Anthropology* 37 (October): 301–15. https://doi.org/10.1146/annurev.anthro.37.081407.085138.
Jolly, Margaret. 1987. "The Forgotten Women: A History of Migrant Labour and Gender Relations in Vanuatu." *Oceania* 58, no. 2 (December): 119–39. https://doi.org/10.1002/j.1834-4461.1987.tb02265.x.
– 1991. "'To Save the Girls for Brighter and Better Lives': Presbyterian Missions and Women in the South of Vanuatu: 1848–1870." *Journal of Pacific History* 26, no. 1 (June): 27–48. https://doi.org/10.1080/00223349108572645.
– 1996. "'Woman Ikat Raet Long Human Raet O No?' Women's Rights, Human Rights and Domestic Violence in Vanuatu." In "The World Upside Down: Feminisms in the Antipodes," edited by Anne Curthoys. Special issue, *Feminist Review* 52 (Spring): 169–90. https://doi.org/10.2307/1395780.
– 1997. "Women-Nation-State in Vanuatu: Women as Signs and Subjects in the Discourses of Kastom, Modernity and Christianity." In *Narratives of Nation in the South Pacific*, edited by Otto Ton and Nicolas Thomas, 133–62. Amsterdam: Harwood.
– 1998. "Other Mothers: Maternal 'Insouciance' and the Depopulation Debate in Fiji and Vanuatu, 1890–1930." In Ram and Jolly 1998, 177–212. Cambridge: Cambridge University Press.
– 2001. "Infertile States: Person and Collectivity, Region and Nation in the Rhetoric of Pacific Population." In *Borders of Being: Citizenship, Fertility and Sexuality in Asia and the Pacific*, edited by Margaret Jolly and Kalpana Ram, 262–306. Ann Arbor: University of Michigan Press.

– 2002. "From Darkness to Light? Epidemiologies and Ethnographies of Motherhood in Vanuatu." In Lukere and Jolly 2002, 148–77.
– 2015. "Braed Praes in Vanuatu: Both Gifts and Commodities?" In "Gender and Person," edited by John P. Taylor and Rachel Morgain. Special issue, *Oceania* 85, no. 1 (March): 63–78. https://doi.org/10.1002/ocea.5074.
Jordan, Brigitte. 1993. *Birth in Four Cultures: A Crosscultural Investigation of Childbirth in Yucatan, Holland, Sweden and the United States*. 4th ed. Long Grove, IL: Waveland Press.
– 1997. "Authoritative Knowledge and Its Construction." In *Childbirth and Authoritative Knowledge: Cross-Cultural Perspectives*, edited by Robbie E. Davis-Floyd and Carolyn Sargent, 55–79. Berkeley: University of California Press.
Kildea, Sue, and Wardaguga, Molly. 2009. "Childbirth in Australia: Aboriginal and Torres Strait Islander Women." In *Childbirth across Cultures: Ideas and Practices of Pregnancy, Childbirth and the Postpartum*, edited by Helain Selin, 275–86. New York: Springer. https://doi.org/10.1007/978-90-481-2599-9_26.
Klein, Naomi. 2020. "Coronavirus Capitalism – and How to Beat It." *The Intercept*, 16 March 2020. YouTube video, 8:48. https://youtu.be/niwNTI9Nqd8?t=19.
Kligman, Gail. 1998. *The Politics of Duplicity: Controlling Reproduction in Ceausescu's Romania*. Berkeley: University of California Press.
Krause, Elizabeth L. 2005. *A Crisis of Births: Population Politics and Family-Making in Italy*. Belmont, CA: Wadsworth.
Krause, Elizabeth L., and Silvia De Zordo. 2012. "Introduction. Ethnography and Biopolitics: Tracing 'Rationalities' of Reproduction across the North–South Divide." In "Irrational Reproduction: New Intersections of Politics, Gender, Race, and Class Across the North-South Divide," edited by Elizabeth L. Krause and Silvia De Zordo. Special issue, *Anthropology & Medicine* 19, no. 2 (August): 137–51. https://doi.org/10.1080/13648470.2012.675050.
Lambek, Michael. 2013. "Kinship, Modernity, and the Immodern." In McKinnon and Cannell 2013, 252–71.
– 2015. *The Ethical Condition: Essays on Action, Person, and Value*. Chicago: University of Chicago Press.
Latour, Bruno. 1993. *We Have Never Been Modern*. Cambridge, MA: Harvard University Press.
Law, John. 2009. "Seeing Like a Survey." In "Narrative, Numbers and Socio-Cultural Change," edited by Shinobu Majima and Niamh Moore. Special issue, *Cultural Sociology* 3, no. 2 (July): 239–56. https://doi.org/10.1177%2F1749975509105533.
Lederman, Rena. 1986. *What Gifts Engender: Social Relations and Politics in Mendi, Highland Papua*. Cambridge: Cambridge University Press.

Li, Tania Murray. 2007. "Governmentality." *Anthropologica* 49 (2): 275–81. https://www.jstor.org/stable/25605363.
Liamputtong, Pranee, and Somsri Kitisriworapan. 2014. "Authoritative Knowledge, Folk Knowledge, and Antenatal Care in Contemporary Northern Thailand." In *Contemporary Socio-Cultural and Political Perspectives in Thailand*, edited by Pranee Liamputtong, 465–86. Dordrecht, NL: Springer.
Lin, Justin Yifu. 2012. "Youth Bulge: A Demographic Dividend or a Demographic Bomb in Developing Countries?" *Let's Talk Development* (blog), World Bank, 5 January 2012. https://blogs.worldbank.org/developmenttalk/youth-bulge-a-demographic-dividend-or-a-demographic-bomb-in-developing-countries.
Lind, Craig. 2014. "Why the Future Is Selfish and Could Kill: Contraception and the Future of Paama." In *Pacific Futures: Projects, Politics and Interests*, edited by Will Rollason, 71–95. New York: Berghahn Books.
Lindstrom, Lamont. 1990. *Knowledge and Power in a South Pacific Society*. Washington, DC: Smithsonian Institution Press.
– 2017. "Respek and Other Urban Vila Keywords." In "Urbanisation en Mélanésie," edited by Lamont Lindstrom and Christine Jourdan. Special issue, *Journal de la Société des Océanistes* 144–5:23–36. https://doi.org/10.4000/jso.7849.
Livingston, Julie. 2012. *Improvising Medicine: An African Oncology Ward in an Emerging Cancer Epidemic*. Durham, NC: Duke University Press.
Lukere, Vicki. 2002. "Native Obstetric Nursing in Fiji." In Lukere and Jolly 2002, 100–24.
Lukere, Vicki, and Margaret Jolly, eds. 2002. *Birthing in the Pacific: Beyond Tradition and Modernity?* Honolulu: University of Hawai'i Press.
MacDonald, Margaret. 2019. "The Image World of Maternal Mortality: Visual Economies of Hope and Aspiration in the Global Campaigns to Reduce Maternal Mortality." *Humanity: An International Journal of Human Rights, Humanitarianism, and Development* 10, no. 2 (Summer): 263–85. https://doi.org/10.1353/hum.2019.0013.
Mahmood, Saba. 2001. "Feminist Theory, Embodiment, and the Docile Agent: Some Reflections on the Egyptian Islamic Revival." *Cultural Anthropology* 16, no. 2 (May): 202–36. https://doi.org/10.1525/can.2001.16.2.202.
Malinowski, Bronislaw. 1916. "Baloma: The Spirits of the Dead in the Trobriand Islands." *Journal of the Royal Anthropological Institute of Great Britain and Ireland* 46 (July–December): 353–430. https://doi.org/10.2307/2843398.
Manderson, Lenore. 1996. *Sickness and the State: Health and Illness in Colonial Malaya, 1870–1940*. Cambridge: Cambridge University Press.
Maternity Worldwide. n.d. "Millennium Development Goal 5 – Results." Maternity Worldwide: Saving Lives in Childbirth. Accessed 20 July 2022.

https://www.maternityworldwide.org/the-issues/achieving-mdg-5-the-facts/.
McArthur, Norma. 1961. *Introducing Population Statistics*. Melbourne: Oxford University Press.
– 1967a. *Island Populations of the Pacific*. Canberra: Australian National University Press.
McArthur, Norma, and John Yaxley. 1968. *Condominium of the New Hebrides: A Report on the First Census of the Population 1967*. Sydney: V.C.N. Blight, Government Printer.
McArthy, Joe. 2020. "Tiny Island Nation of Vanuatu Pledges Millions to Help Australian Firefighters." *Global Citizen*, 8 January 2020. https://www.globalcitizen.org/en/content/vanuatu-pledges-20-million-australia-fires/.
McCann, Carole. 2017. *Figuring the Population Bomb: Gender and Demography in the Mid-Twentieth Century*. Seattle: University of Washington Press.
McDonnell, Siobhan. 2013. "Exploring the Cultural Power of Land and Law in Vanuatu: Law as a Performance that Creates Meaning and Identities." *Intersections: Gender and Sexuality in Asia and the Pacific* 33 (December). http://intersections.anu.edu.au/issue33/mcdonnell.htm.
– 2015. "The Land Will Eat You: Land and Sorcery in North Efate, Vanuatu." In *Talking It Through: Responses to Sorcery and Witchcraft Beliefs and Practices in Melanesia*, edited by Miranda Forsyth and Richard Eves, 137–60. Canberra: ANU Press. https://doi.org/10.22459/TIT.05.2015.
– 2016. "'My Land, My Life': Power, Property and Identity in Land Transformations in Vanuatu." PhD diss., Australia National University.
McKay, Ramah. 2018. *Medicine in the Meantime: The Work of Care in Mozambique*. Durham, NC: Duke University Press.
McKelvie, Stephanie, Basil Leodoro, Thomas Sala, Thach Tran, and Jane Fisher. 2020. "Prevalence, Patterns, and Determinants of Intimate Partner Violence Experienced by Women Who Are Pregnant in Sanma Province, Vanuatu." In "Bullying and Its Linkage to Other Forms of Violence," edited by Jun Sung Hong, Dorothy L. Espelage and Jamie M. Ostrov. Special issue, *Journal of Interpersonal Violence* 37, nos. 9–10 (May): NP7632–53. https://doi.org/10.1177/0886260520969235.
McKinnon, Susan. 2019. "Reading the Contested Forms of Nation through the Contested Forms of Kinship and Marriage." In *The Cambridge Handbook of Kinship*, edited by Sandra Bamford, 605–28. Cambridge: Cambridge University Press. https://doi.org/10.1017/9781139644938.026.
McKinnon, Susan, and Fenella Cannell, eds. 2013. *Vital Relations: Modernity and the Persistent Life of Kinship*. Santa Fe, NM: School for Advanced Research Press.

McMenamin, Dorothy. 2009. "Leprosy and Stigma in the South Pacific: Camaraderie in Isolation." Master's thesis, University of Canterbury. https://doi.org/10.26021/4476.

McPherson, Naomi M. 2007. "Women, Childbirth and Change in West New Britain, Papua New Guinea." In *Reproduction, Childbearing and Motherhood*, edited by Pranee Liamputtong, 127–41. New York: Nova Science.

Merry, Sally Engle. 2016. *The Seductions of Quantification: Measuring Human Rights, Gender Violence, and Sex Trafficking*. Chicago: University of Chicago Press.

Meyerhoff, Miriam. 2002. "A Vanishing Act: Tonkinese Migrant Labour in Vanuatu in the Early 20th Century." *Journal of Pacific History* 37, no. 1 (June): 45–56. https://doi.org/10.1080/00223340220139261.

Miller, J. Graham. 1978. *LIVE: A History of Church Planting in the New Hebrides.* Book 1, *To 1880.* Sydney: Committees on Christian Education and Overseas Missions, General Assembly of the Presbyterian Church of Australia.

– 1986. *LIVE: A History of Church Planting in the Republic of Vanuatu.* Book 4, *1881–1920.* Sydney: Committees on Christian Education and Overseas Missions, General Assembly of the Presbyterian Church of Australia.

Mitchell, Jean. 2011a. "Engaging Feminist Anthropology in Vanuatu: Local Knowledge and Universal Claims." In "Feminism and Anthropology," edited by Robin Whitaker and Pamela J. Downe. Special issue, *Anthropology in Action* 18, no. 1 (March): 29–41. https://doi.org/10.3167/aia.2011.180105.

– 2011b. "'Operation Restore Public Hope': Youth and the Magic of Modernity in Vanuatu." *Oceania* 81, no. 1 (March): 36–50. https://doi.org/10.1002/j.1834-4461.2011.tb00092.x.

– 2021. "'Awakening the Stones': The Nieri Performance, Gardens and Regeneration in Tanna, Vanuatu." In "The Art of Gardens: Views from Melanesia and Amazonia," edited by Jean Mitchell and Lissant Bolton. Anthropological Forum 31, no. 4 (October): 433–49. https://doi.org/10.1080/00664677.2021.2004878.

Mitchell, Timothy. 2001. *Rule of Experts: Egypt, Techno Politics, Modernity.* Berkeley: University of California Press.

MNCC (Malvatumauri National Council of Chiefs). 2012. *Alternative Indicators of Well-Being for Melanesia: Vanuatu Pilot Study Report*. Port Vila: Vanuatu National Statistics Office.

MNCC (Malvatumauri National Council of Chiefs) and VNSO (Vanuatu National Statistics Office). 2012. *Well-Being in Vanuatu.* Written by Jamie Tanguay. VKS Productions. YouTube video, 20:00. https://youtu.be/jtnLl1Jp0K0.

Mol, Annemarie, Ingunn Moser, and Jeannette Pols. 2010. "Care: Putting Practice into Theory." In *Care in Practice: On Tinkering in Clinics, Homes and*

Farms, edited by Annemarie Mol, Ingunn Moser, and Jeannette Pols, 7–26. Bielefeld, Germany: Transcript Verlag.
Morgan, Lynn M. 2019. "Reproductive Governance, Redux." *Medical Anthropology: Cross-Cultural Studies in Health and Illness* 38, no. 2 (February): 113–17. https://doi.org/10.1080/01459740.2018.1555829.
Morgan, Lynn M., and Elizabeth F.S. Roberts. 2012. "Reproductive Governance in Latin America." In "Irrational Reproduction: New Intersections of Politics, Gender, Race, and Class Across the North-South Divide," edited by Elizabeth L. Krause and Silvia De Zordo. Special issue, *Anthropology & Medicine* 19, no. 2 (August): 241–54. https://doi.org/10.1080/13648470.2012.675046.
Murphy, Michelle. 2013. "Distributed Reproduction, Chemical Violence, and Latency." In "Life (Un)Ltd: Feminism, Bioscience, Race," edited by Rachel C. Lee. Special issue, *The Scholar & Feminist Online* 11, no. 3 (Summer). https://sfonline.barnard.edu/life-un-ltd-feminism-bioscience-race/distributed-reproduction-chemical-violence-and-latency/.
– 2015a. "Reproduction." In *Marxism and Feminism*, edited by Shahrzad Mojab, 287–304. New York: Zed Books.
– 2015b. "Unsettling Care: Troubling Transnational Itineraries of Care in Feminist Health Practices." In "The Politics of Care in Technoscience," edited by Ana Viseu, Natasha Myers, Aryn Martin, and Lucy Suchman. Special issue, *Social Studies of Science* 45, no. 5 (October): 717–37. https://doi.org/10.1177%2F0306312715589136.
– 2017. *The Economization of Life*. Durham, NC: Duke University Press.
National Planning and Statistics Office. 1983. Report on the Census of Population, 1979. Port Vila: National Planning and Statistics Office.
Naupa, Anna. 2017. "Making the Invisible Seen: Putting Women's Rights on Vanuatu's Land Reform Agenda." In *Kastom, Property and Ideology: Land Transformations in Melanesia*, edited by Siobhan McDonnell, Matthew Allen, and Colin Filer, 305–26. Canberra: ANU Press.
Obituaries Australia. 1984. "McArthur, Norma (?–1984)." Obituaries Australia, National Centre of Biography, Australian National University. Accessed 8 November 2019. https://oa.anu.edu.au/obituary/mcarthur-norma-677/text678.
Olszynko-Gryn, Jesse. 2014. "Laparoscopy as a Technology of Population Control: A Use-Centered History of Surgical Sterilization." In *A World of Populations: The Production, Transfer, and Application of Demographic Knowledge in the Twentieth Century in Transnational Perspective*, edited by Heinrich Hartmann and Corinna R. Unger, 147–77. New York: Berghahn Books.
Pacific Community. n.d. "Mary Kalsrap." South Pacific Development. Accessed 21 April 2020. https://www.spc.int/70-inspiring-pacific-women/mary-kalsrap/.

Pacific Institute of Public Policy. 2011. "Youthquake: Will Melanesian Democracy Be Sunk by Demography?" Discussion paper no. 17. Port Vila: Pacific Institute of Public Policy. https://actnowpng.org/sites/default/files/Youthquake- will Melanesian democracy be sunk by demography.pdf.

Pacific Islands Monthly. 1967. "Big Bay Story Brings Land Tiffs into Open." *Pacific Islands Monthly* 38, no. 5 (May): 31–2.

Pala'amo, Alesana. 2019. "Pastoral Counselling in a Changing Samoa: Development, Christianity and Relationality." *Sites* 16 (1): 95–108. https://doi.org/10.11157/sites-id442.

Philibert, Jean-Marc. 1988. "Women's Work: A Case Study of Proletarianization of Peri-Urban Villagers in Vanuatu." *Oceania* 58, no. 3 (March): 161–75. https://doi.org/10.1002/j.1834-4461.1988.tb02270.x.

Population and Development Program. 2006. "Ten Reasons to Rethink Overpopulation." Population and Development Program, Hampshire College. https://sites.hampshire.edu/popdev/10-reasons-to-rethink-overpopulation/.

Population Knowledge Network, eds. 2016. *Twentieth Century Population Thinking: A Critical Reader of Primary and Secondary Sources*. London: Routledge.

Porter, Theodore M. (1995) 2020. *Trust in Numbers: The Pursuit of Objectivity in Science and Public Life*. Princeton, NJ: Princeton University Press.

PPB (Pacific Private Bank). n.d. "About Us." Pacific Private Bank. Accessed 21 April 2020. https://pacificprivatebank.com/about-us/.

Ram, Kalpana, and Margaret Jolly, eds. 1998. *Maternities and Modernities: Colonial and Postcolonial Experiences in Asia and the Pacific*. Cambridge: Cambridge University Press.

– 1992. "Family and Class in Contemporary America: Notes Toward an Understanding of Ideology." In Thorne and Yalom, 25–39.

Rawlings, Gregory E. 1999a. "Foundations of Urbanization: Port Vila Town and Pango Village, Vanuatu." *Oceania* 70, no. 1 (September): 72–86. https://doi.org/10.1002/j.1834-4461.1999.tb02990.x.

– 1999b. "Villages, Islands and Tax Havens: The Global/Local Implications of a Financial Entrepot in Vanuatu." *Canberra Anthropology* 22, no. 2 (October): 37–50. https://doi.org/10.1080/03149099909508347.

– 2002. "'Once There Was a Garden, Now There Is a Swimming Pool": Inequality, Labour and Land in Pango, a Peri-urban Village in Vanuatu." PhD diss., Australian National University.

– 2004. "Laws, Liquidity and Eurobonds: The Making of the Vanuatu Tax Haven." *Journal of Pacific History* 39, no. 3 (December): 325–41. https://doi.org/10.1080/0022334042000290388.

– 2019. "Stateless Persons, Eligible Citizens and protected Places: The British Nationality Act in Vanuatu." *Twentieth Century British History* 30, no. 1 (March): 53–80. https://doi.org/10.1093/tcbh/hwy011.

Razak, Iskhandar, and Graeme Hugo. 2012. "Population Pressure Could Cause 'Melanesian Spring': Demographer." *ABC Radio Australia*. Accessed 8 February 2013. http://www.radioaustralia.net.au/international/radio/program/pacific-beat/population-pressure-could-cause-melanesian-spring-demographer/1025884.

Regenvanu, Ralph. 2021. "Untitled Status Update." Facebook, 13 July 2021. https://www.facebook.com/profile/701818847/search/?q=well-being.

Reinecke, Christiane. 2016. "Population in Space: Migration, Geopolitics, and Urbanization." In Population Knowledge Network 2016, 90–114.

Rio, Knut. 2002. "The Sorcerer as an Absented Third Person: Formations of Fear and Anger in Vanuatu." In "Beyond Rationalism: Rethinking Magic, Witchcraft and Sorcery," edited by Bruce Kapferer. Special issue, *Social Analysis* 46, no. 3 (Fall): 129–54. https://doi.org/10.3167/015597702782409284.

– 2007. "Exposer la vie après la mort : du bon usage social des prestations mortuaires au Vanuatu." *Journal de la Société des Océanistes* 124 (1): 67–81. https://doi.org/10.4000/jso.778.

Riou, Virginie. 2010. "Trajectoires pseudo-coloniales. Les Français du condominium franco-anglais des ex Nouvelles-Hébrides (Vanuatu) de la fin du XIXe siècle à l'entre deux guerres." PhD diss., École des hautes études en sciences sociales.

Rivers, William H.R. 1922. *Essays on the Depopulation of Melanesia*. Cambridge: Cambridge University Press.

– 2015. "Reproduction and Cultural Anthropology." In International Encyclopedia of the Social and Behavioral Sciences, edited by James D. Wright, 2nd ed., vol. 20, 450–56. Amsterdam: Elsevier. https://doi.org/10.1016/B978-0-08-097086-8.12239-1.

Robertson, Thomas. 2016. "Natural Resources, Environment and Population." In Population Knowledge Network, 210–35.

Robinson-Drawbridge, Benjamin. 2020. "As If a Bomb Went off: Vanuatu's Pentecost devastated by Cyclone Harold." *Radio New Zealand International*, 15 April 2020. https://www.rnz.co.nz/international/pacific-news/414299/as-if-it-was-bombed-vanuatu-s-pentecost-devastated-by-cyclone-harold.

Rodman, Margaret Critchlow. 1987. *Masters of Tradition: Consequences of Customary Land Tenure in Longana, Vanuatu*. Vancouver: University of British Columbia Press.

– 2001. *Houses Far from Home: British Colonial Space in the New Hebrides*. Honolulu: University of Hawai'i Press.

– 2003. "The Heart in the Archives: Colonial Contestation of Desire and Fear in the New Hebrides, 1993." *Journal of Pacific History* 38, no. 3 (December): 291–312. https://doi.org/10.1080/0022334032000154056.

Rodman, Margaret, Daniela Kraemer, Lissant Bolton, and Jean Tarisesei, eds. 2007. *House-Girls Remember: Domestic Workers in Vanuatu*. Honolulu: University of Hawai'i Press.

Rodríguez-Muñiz, Michael. 2017. "Cultivating Consent: Nonstate Leaders and the Orchestration of State Legibility." *American Journal of Sociology* 123, no. 2 (September): 385–425. https://doi.org/10.1086/693045.

Rousseau, Benedicta, and John P. Taylor. 2012. "Kastom Ekonomi and the Subject of Self-Reliance: Differentiating Development in Vanuatu." In *Differentiating Development: Beyond an Anthropology of Critique*, edited by Soumhya Venkatesan and Thomas Yarrow, 169–86. New York: Berghahn Books.

Sanger, Margaret, ed. 1927. *World Population Conference Proceedings*. London: Edward Arnold.

Sasser, Jade. 2014. "From Darkness into Light: Race, Population, and Environmental Advocacy." *Antipode* 46, no. 5 (November): 1240–57. https://doi.org/10.1111/anti.12029.

– 2018. *On Infertile Ground: Population Control and Women's Rights in the Era of Climate Change*. New York: New York University Press.

Scherz, China. 2018. "Enduring the Awkward Embrace: Ontology and Ethical Work in a Ugandan Convent." *American Anthropologist* 120, no. 1 (March): 102–12. https://doi.org/10.1111/aman.12968.

Scott, David. 1995. "Colonial Governmentality." *Social Text* 43 (Autumn): 191–220. https://doi.org/10.2307/466631.

– 2004. *Conscripts of Modernity: The Tragedy of Colonial Enlightenment*. Durham, NC: Duke University Press.

Scott, James. 1998. *Seeing Like a State: How Certain Schemes to Improve the Human Condition Have Failed*. New Haven, CT: Yale University Press.

Selwyn Foundation. 2017. "Betty Pyatt Apartments." Selwyn Foundation. Accessed 12 April 2020. https://www.selwynfoundation.org.nz/villages/independent-living/selwyn-village/living-options/betty-pyatt-apartments/.

Sénat Français. n.d. "Le Vanuatu: Survivance de la Francophonie dans un archipel du Pacifique du Sud." Sénat.fr. Accessed 9 March 2020. https://www.senat.fr/ga/ga33/ga332.html.

Servy, Alice. 2018. "L'approche « Abstinence, Be Faithful, Use a Condom » au Vanuatu : Traduire une politique d'éducation à la sexualité globalisée." In "Variations," *Autrepart* 86 (2): 43–59. https://doi.org/10.3917/autr.086.0043.

– 2020. "'We've Paid Your Vagina to Make Children!': Bridewealth and Women's Marital and Reproductive Autonomy in Port-Vila, Vanuatu." In "Bridewealth and the Autonomy of Women," edited by Christine Jourdan and Karen Sykes. *Oceania* 90, no. 3 (November): 292–308. https://doi.org/10.1002/ocea.5280.

Shever, Elana. 2013. "'I Am a Petroleum Product': Making Kinship Work on the Patagonian Frontier." In McKinnon and Cannell, 99–119.

Simpson, Audra. 2014. *Mohawk Interruptus: Political Life across the Borders of Settler States*. Durham, NC: Duke University Press.

Solway, Richard. 1990. *Demography and Degeneration: Eugenics and the Declining Birthrate in Twentieth-Century Britain*. Durham: University of North Carolina Press.

Speiser, Felix. 1913. *Two Years with the Natives in the Western Pacific*. London: Mills and Boon.

– (1923) 1996. *Ethnology of Vanuatu: An Early 20th Century Study*. Bathurst, UK: Crawford House.

Spriggs, Matthew. 1997. *The Island Melanesians*. Oxford: Blackwell.

– 2007. "Population in a Vegetable Kingdom. Aneityum Island (Vanuatu) At European Contact in 1830." In *The Growth and Collapse of Pacific Island Societies*, edited by Patrick Kirch and Jean-Louis Rallu, 278–305. Honolulu: University Press of Hawai'i.

Smith, Thomas Richard. 1972. *South Pacific Commission: An Analysis after Twenty-Five Years*. Wellington: Price Milburn for the New Zealand Institute of International Affairs.

Southern Cross. 1950. *Medical Work in Melanesia: The Solomon Islands and the New Hebrides*. Southern Cross Booklet no. 7. Oxford: Church Army Press. Project Canterbury. http://anglicanhistory.org/oceania/sx_booklet7.html.

Stevens, Kate. 2017. "'The Law of the New Hebrides Is the Protector of Their Lawlessness': Justice, Race and Colonial Rivalry in the Early Anglo-French Condominium." *Law and History Review* 35, no. 3 (August): 595–620. https://doi.org/10.1017/S0738248017000293.

Stevenson, Lisa. 2014. *Life Beside Itself: Imagining Care in the Canadian Arctic*. Berkeley: University of California Press.

Stoler, Ann Laura. 1995. *Race and the Education of Desire: Foucault's History of Sexuality and the Colonial Order of Things*. Durham, NC: Duke University Press.

– 2002a. *Carnal Knowledge and Imperial Power: Race and the Intimate in Colonial Rule*. Berkeley: University of California Press.

– 2002b. "Colonial Archives and the Arts of Governance." *Archival Science* 2, nos. 1–2 (March): 87–109. https://doi.org/10.1007/BF02435632.

– 2010. *Along the Archival Grain: Epistemic Anxieties and Colonial Common Sense*. Princeton, NJ: Princeton University Press.

Strathern, Marilyn. 1988. *The Gender of the Gift: Problems with Women and Problems with Society in Melanesia*. Berkeley: University of California Press.

Street, Alice. 2014. *Biomedicine in an Unstable Place: Infrastructure and Personhood in a Papuan New Guinean Hospital*. Durham, NC: Duke University Press.

– 2016. "Making People Care." *The Lancet* 387, no. 10016 (23 January 2016): 333–4. https://doi.org/10.1016/S0140-6736(16)00119-7.

– 2019. "Health, Institutions and Governance in Melanesia." In *The Melanesian World*, edited by Eric Hirsch and Will Rollason, 300–14. London: Routledge.

Tan, Yvette. 2020. "Cyclone Harold and Coronavirus: Pacific Islands Face Battle on Two Fronts." *BBC News*, 15 April 2020. https://www.bbc.com/news/world-asia-52268119.

Tanguay, Jamie. 2014. "Alternative Indicators of Well-Being for Melanesia: Changing the Way Progress Is Measured in the South Pacific." *Devpolicy* (blog), 21 May 2014. https://www.devpolicy.org/alternative-indicators-of-well-being-for-melanesia-changing-the-way-progress-is-measured-in-the-south-pacific-20140521/.

– 2015. "Alternative Indicators of Wellbeing for Melanesia: Cultural Values Driving Public Policy." In *Making Culture Count: The Politics of Cultural Measurement*, edited by Lachlan MacDowall, Marnie Badham, Emma Blomkamp, and Kim Dunphy, 162–72. London: Palgrave Macmillan. https://doi.org/10.1007/978-1-137-46458-3.

Tarlo, Emma. 2003. *Unsettling Memories: Narratives of Emergency in Delhi*. Durham, NC: Duke University Press.

Teaiwa, Katerina. 2015. *Consuming Ocean Island: Stories of People and Phosphate from Banaba*. Bloomington: University of Indiana Press.

Teman, Elly. 2003. "'Knowing' the Surrogate Body in Israel." In *Surrogate Motherhood: International Perspectives*, edited by Rachel Cook, Shelley Day-Sclater, and Felicity Kaganas, 261–79. Portland, OR: Hart Publishing.

Thorne, Barrie, and Marilyn Yalom, eds. 1992. *Rethinking the Family: Some Feminist Questions*, 2nd ed. Boston: Northeastern University Press.

Ticktin, Miriam. 2006. "Where Ethics and Politics Meet: The Violence of Humanitarianism in France." *American Ethnologist* 33, no. 1 (February): 33–49. https://doi.org/10.1525/ae.2006.33.1.33.

Tokona, Mavuku. 2019. "VCH Maternity Ward Overhaul Long-Overdue." *Vanuatu Daily Post*, 5 March 2020. https://dailypost.vu/news/vch-maternity-ward-overhaul-long-overdue/article_c59a50ea-2713-5cbe-bdb6-c3e5a1c248e1.html.

Tonkinson, Robert. 1982. "National Identity and the Problem of Kastom in Vanuatu." *Mankind* 13, no. 4 (August): 306–15. https://doi.org/10.1111/j.1835-9310.1982.tb00996.x.

Tuidraki, Peni. 1957. "New Hebrides Midwife." *Native Medical Practitioner* 3 (2): 488–9.

UNFPA (United Nations Population Fund). 2012. "Vanuatu Country Profile." UNFPA. Accessed 8 February 2013. http://www.unfpa.org/public/_ns/YWRleGVyby1mZC1wb3J0bGV0OjpkZXhlcm8tZmQtcG9ydGxldDo6LTYyYTZhNDI5OjEyOTk2MDgxYzg3Oi03ZmYxfGVhY3Rpb249MT12aWV3RGF0YVZpZXc_/cache/bypass/appid/310297_1/home/sitemap/countries.

– 2014. Population and Development Profiles: Pacific Island Countries. Suva, FJ: UNFPA Pacific. https://pacific.unfpa.org/sites/default/files

/pub-pdf/web__140414_UNFPAPopulationandDevelopmentProfiles-PacificSub-RegionExtendedv1LRv2_0.pdf.
– 2016. *State of World Population 2016*. New York: UNFPA.
– 2019. *State of Pacific Youth Report 2017*. Suva, FJ: UNFPA Pacific.
UNICEF. 2013. "Vanuatu Statistics." UNICEF. Accessed 26 June 2020. https://www.unicef.org/infobycountry/vanuatu_statistics.html.
Van Hollen, Cecilia. 2003. *Birth on the Threshold: Childbirth and Modernity in South India*. Berkeley: University of California Press.
Van Trease, Howard. 1987. *The Politics of Land in Vanuatu Suva*. Suva, FJ: University of the South Pacific Press.
Vanuatu Daily Post. 2020a. "Government to Invest in Shepherd Islands through NDMO and FSAC." *Vanuatu Daily Post*, 23 May 2020. https://dailypost.vu/news/government-to-invest-in-shepherd-islands-through-ndmo-and-fsac/article_df2ac640-9e0a-11ea-814c-1f5b707e6b00.html.
– 2020b. "Over VT7 Million Donated to VCH Maternity Ward by Pacific Private Bank." *Vanuatu Daily Post*, 18 April 2020. https://dailypost.vu/news/over-vt7-million-donated-to-vch-maternity-ward-by-pacific-private-bank/article_47705c4a-85e9-11ea-92e9-2780018056ec.html.
Vanuatu Department of Laefstok. 2020. "These 'Jaw-Dropping' Footage Shows Piles and Piles of Yams." Facebook, 27 May 2020. https://www.facebook.com/groups/126303117967224/permalink/641809973083200/.
Vanuatu Department of Strategic Policy, Planning and Aid Coordination. 2011. *National Population Policy 2011–2020*. Port Vila: Government of Vanuatu.
Vanuatu Ministry of Health, VNSO (Vanuatu National Statistics Office), and SPC (Secretariat of the Pacific Community). 2014. *Demographic and Health Survey 2013: Final Report*. Port Vila: VNSO.
VNSO (Vanuatu National Statistics Office). 1993. Vanuatu National Population Census 1989: Demographic and Migration Analysis. Port Villa: VNSO. https://purl.org/spc/digilib/doc/25wd8.
– 2001. National Population Census 1989: Demographic Analysis Report. Port Vila: VNSO.
– 2010. National Population and Housing Census 2009: Summary Release. Port Vila: Government of Vanuatu. https://vnso.gov.vu/index.php/en/census-and-surveys/census/2009-census#census-archived-reports.
– 2016. *Post Pam Vanuatu Population and Housing Mini-Census Report*, vol 1. Port Vila: Government of Vanuatu.
– 2021a. *Well-Being in Vanuatu: 2019–2020 National Sustainable Development Plan (NSDP) Baseline Survey*. Port Vila: VNSO.
– 2021b. *2020 National Population and Housing Census: Basic Tables Report*, vol. 1. Port Villa: VNSO. https://vnso.gov.vu/images/Pictures/Census/2020_census/Census_Volume_1/2020NPHC_Volume_1.pdf.

VNSO (Vanuatu National Statistics Office), Andy Calo, Richard Curtain, Simil Johnson, Benuel Lenge, Melanie Nalau, and Jimmy Tamkela. 2012. *Youth Monograph: Young People in Vanuatu*. Port Vila: VNSO.

VNSO (Vanuatu National Statistics Office), Christensen Fund, SPC (Secretariat of the Pacific Community), and Vanuatu Cultural Centre. 2010. *Alternative Indicators of Well-Being for Melanesia: Changing the Way Progress Is Measured in the South Pacific*. Port Vila: VNSO.

Wardlow, Holly. 2006. *Wayward Women: Sexuality and Agency in a New Guinea Society*. Berkeley: University of California Press.

– 2020. *Fencing in AIDS: Gender, Vulnerability, and Care in Papua New Guinea*. Berkeley: University of California Press. https://doi.org/10.1525/luminos.94.

Ware, Helen. 2005. "Demography, Migration and Conflict in the Pacific." In "Demography of Conflict and Violence," edited by Helge Brunborg and Henrik Urdal. Special issue, *Journal of Peace Research* 42, no. 4 (July): 435–54. https://doi.org/10.1177/0022343305054090.

Waring, Marilyn. 1988. *If Women Counted: A New Feminist Economics*. New York: Harper and Row.

Weiner, Annette B. 1977. "Trobriand Descent: Female/Male Domains." *Ethos* 5, no. 1 (Spring): 54–70. https://doi.org/10.1525/eth.1977.5.1.02a00050.

– 1980. "Reproduction: A Replacement for Reciprocity." *American Ethnologist* 7, no. 1 (February): 71–85. https://doi.org/10.1525/ae.1980.7.1.02a00050.

WHO (World Health Organization). 2020. "Unmet Need for Family Planning (%)." Global Health Observatory, World Health Organization. Accessed 4 June 2020. https://www.who.int/data/gho/indicator-metadata-registry/imr-details/3619.

Wickramasinghe, Nira. 2015. "Colonial Governmentality and the Political Thinking through '1931' in the Crown Colony of Ceylon/Sri Lanka." In "Inventer les sciences sociales postoccidentales," edited by Laurence Roulleau-Berger. Special issue, *Socio* 5:99–114. https://doi.org/10.4000/socio.1921.

Widmer, Alexandra. 2008. "The Effects of Elusive Knowledge: Census, Health Laws and Inconsistently Modern Subjects in Early Colonial Vanuatu." *Journal of Legal Anthropology* 1, no. 1 (September): 92–116. https://doi.org/10.3167/jla.2008.010105.

– 2010. "Native Medical Practitioners, Temporality and Nascent Biomedical Citizenship in the New Hebrides." In "At Disciplinary Edges," edited by Elizabeth Mertz and Katherine Bowie. Special supplement issue, *Political and Legal Anthropology Review* 30, no. S1 (May): 57–80. https://doi.org/10.1111/j.1555-2934.2010.01066.x.

– 2012. "Of Field Encounters and Metropolitan Debates: Research and the Making and Meaning of the Melanesian 'Race' during

Demographic Decline." *Paideuma: Mitteilungen zur Kulturkunde* 58:69–93. https://www.jstor.org/stable/23644454.
– 2013. "Seeing Health Like a Colonial State: Assistant Medical Practitioners and Nascent Biomedical Citizenship in the New Hebrides." In *Senses and Citizenships: Embodying Political Life*, edited by Susanna Trnka, Julie Park, and Christine Dureau, 200–20. New York: Routledge.
– 2014. "The Imbalanced Sex Ratio and the High Bride Price: Watermarks of Race in Demography, Census, and the Colonial Regulation of Reproduction." In "Technologies of Belonging," edited by Amade M'charek, Katharina Schramm, and David Skinner. Special issue, *Science, Technology, & Human Values* 39, no. 4 (July): 538–60.
– 2017 "Making People Countable: Analyzing Paper Trails and the Imperial Census." In *Sources and Methods in Histories of Colonialism: Approaching the Imperial Archive*, edited by Kirsty Reid and Fiona Paisley, 100–16. London: Routledge.
Willie, Glenda. 2019. "From 15 to 33 Beds at Maternity Ward." *Vanuatu Daily Post*, 2 November 2019. https://dailypost.vu/news/from-15-to-33-beds-at-maternity-ward/article_65a420c2-fcfd-11e9-81b3-9730bff7808a.html.
World Bank. 2020. "Life Expectancy at Birth, Total (Years) – Vanuatu." World Bank. Accessed 8 June 2022. https://data.worldbank.org/indicator/SP.DYN.LE00.IN?locations=VU.

Unpublished References

New Hebrides British Service (NHBS) materials are held at the Western Pacific Archive, University of Auckland.

Adams, G.C. 1941. Letter to BRC, 1 February 1941. MSS.Archives.2003/1. NHBS 7. Series II. File 3/1, Marriages by Native Customs and Christian Divorce, 1935–50, Box C2453704.

Ballard, C. 1937. Letter to BDA G.C. Adams, 28 August 1937. MSS. Archives.2003/1.NHBS 7. Series II. File 3/1, Marriages by Native Customs and Christian Divorce, 1935–50, Box C2453704.

BDA (British District Agent). 1939. Letter to BRC, 25 July 1939. MSS. Archives.2003/1.NHBS 7. Series II. File 3/1, Marriages by Native Customs and Christian Divorce, 1935–50, Box C2453704.

– 1940a. Letter to BRC, 4 June 1940. MSS.Archives.2003/1.NHBS 7. Series II. File 3/1, Marriages by Native Customs and Christian Divorce, 1935–50, Box C2453704.

– 1940b. Letter to BRC, 6 June 1940. MSS.Archives.2003/1.NHBS 7. Series II. File 3/1, Marriages by Native Customs and Christian Divorce, 1935–50, Box C2453704.

– 1945a. Letter to BRC, Quarterly Report on "Bride Price," 15 April 1945. MSS. Archives.2003/1.NHBS 7. Series II. File 3/7, The Price of Brides, 1945–50, Box C2453704.

– 1945b. Letter to BRC, Quarterly Report on "Bride Price," 16 July 1945. MSS. Archives.2003/1.NHBS 7. Series II. File 3/7, The Price of Brides, 1945–50, Box C2453704.

– 1945c. Letter to BRC, Quarterly Report on "Bride Price," 23 July 1945. MSS. Archives.2003/1.NHBS 7. Series II. File 3/7, The Price of Brides, 1945–50, Box C2453704.

Blandy, Richard. 1930. Letter to BDA James Nicol, 18 March 1930. MSS. Archives.2003/1.NHBS 15. Series I. File 5/1, Pt. 1, Native Court – General, 1916–48, Box C3030550.

– 1944. Memorandum to BDAs, 28 August 1944. MSS.Archives.2003/1.NHBS 7. Series II. File 3/7, The Price of Brides, 1945–50, Box C2453704.
BRC (British Resident Commissioner) Office. n.d. [December 1941?]. Memo to District Agents: "Marriages by Native Customs and Christian Divorce." MSS.Archives.2003/1.NHBS 7. Series II. File 3/1, Marriages by Native Customs and Christian Divorce, 1935–50, Box C2453704.
Brookfield, Harold, Paula Brown, and Marnie Anderson. 1966. "Social Survey Port Vila." MSS.Archives.2003/1.NHBS 18. Series I. File 140/5, Pt. 1, Demography. First Census N.H. (General), 1962–70, Box C3027052.
Crozier, Ron. 1950. Letter to BRC, 4 July 1950. MSS.Archives.2003/1.NHBS 7. Series II. File 3/7, The Price of Brides, 1945–50, Box C2453704.
District Agents. 1965. Extract Taken from the Minutes of British District Agents Meeting, 15 June 1965. MSS.Archives.2003/1.NHBS 18. Series I. File 140/3, Pt. 1, Demography. Population Statistics (General), 1961–71, Box C3027052.
Flaxman, Hubert. BRC Letter to BDA Crozier, 23 May 1950. MSS.Archives.2003/1. NHBS 7. Series II. File 3/7, The Price of Brides, 1945–50, Box C2453704.
Frater, Maurice. 1916. Letter to BRC Merton King, 27 September 1916. MSS. Archives.2003/1.NHBS 17. Series I. File 6, Vital Statistics Land. etc., 1911–16, Box C3030113.
Freeman, Ted. 1964. Letter to BRC, 18 July 1964. MSS.Archives.2003/1.NHBS 18. Series I. File 295/14, Pt. 1, Health Service. P.M.H. Maternity Unit, 1964–6, Box C3027043.
Heard, Margery, and Elizabeth Pyatt. 1966. "Presentation at Medical Advisory Council Meeting," 21 February 1966. Proposed Syllabus for Nursing Training for New Hebridean Nurses. MSS.Archives.2003/1.NHBS 18. Series I. File 295/13 Pt. 1, Medical. Paton Memorial Hospital – General, 1961–9, Box C3027043.
Joy, George. 1932. Letter to BDA Nicol, 7 January 1932. MSS.Archives.2003/1.NHBS 15. Series I. File 5/1, Pt. 1, Native Court – General, 1916–48, Box C3030550.
– 1934a. Letter to FRC, 11 July 1934. MSS.Archives.2003/1.NHBS 15. Series I. File 5/1, Pt. 1, Native Court – General, 1916–48, Box C3030550.
– 1934b. Letter to FRC, 6 August 1934. MSS.Archives.2003/1.NHBS 15. Series I. File 5/1, Pt. 1, Native Court – General, 1916–48, Box C3030550.
– 1938a. Letter to Reverend Paton, 7 January 1938. MSS.Archives.2003/1.NHBS 7. Series II. File 3/1, Marriages by Native Customs and Christian Divorce, 1935–50, Box C2453704.
– 1938b. Letter to Reverend Paton, 17 February 1938. MSS.Archives.2003/1.NHBS 7. Series II. File 3/1, Marriages by Native Customs and Christian Divorce, 1935–50, Box C2453704.
Mackereth, Bruce. 1963. *Godden Memorial Hospital Annual Report 1963*. MSS. Archives.2003/1.NHBS 8. Series III. File F8/3/1, Godden Memorial Hospital: General, 1959–66, Box C3026132.

– 1966. *Godden Memorial Hospital Annual Report 1966*. MSS.Archives.2003/1.NHBS 8. Series III. File F8/3/1, Godden Memorial Hospital: General, 1959–66, Box C3026132.

Mahaffy, Arthur, and Jules Repiquet. 1912. A Joint Regulation, 20 December 1912. "Native Court, General." MSS.Archives.2003/1.NHBS 15. Series I. File 5/1, Pt. 1, Native Court – General, 1916–48, Box C3030550.

Martin, Ruth. 1961. Letter to BRC, 18 April 1961. MSS.Archives.2003/1.NHBS 18. Series I. File 140/3 Pt. 1, Demography. Population Statistics (General), 1961–71, Box C3027052.

McArthur, Norma. 1965. Letter to BRC Colin Allan, Outline of Census Procedure, 13 January 1965. MSS.Archives.2003/1.NHBS 18. Series I. File 140/5, Pt. 1, Demography. First Census N.H. (General), 1962–70, Box C3027052.

– 1967b. Letter to Census Takers. MSS.Archives.2003/1.NHBS 18. Series I. File 140/10, Vol. 2, Census (N.H. Enumerators and Supervisors, 1966–7), Box C3026840.

Medical Advisory Committee. 1967. Minutes of Medical Advisory Committee, 11 April 1967. MSS.Archives.2003/1.NHBS 18. Series I. File 295/3, Pt. 1, Health Service. British Medical Advisory Committee, 1965–70, Box C2453581.

Mills, A.R. 1954. "General Review of Medical Services in the Anglo-French Condominium of the New Hebrides for the Year 1954." Unpublished manuscript in author's possession.

Ministrant. 1966. Ministry of Overseas Development to BRC, 12 April 1966. MSS.Archives.2003/1.NHBS 18. Series I. File 140/3, Pt. 1, Demography. Population Statistics (General), 1961–71, Box C3027052.

Native Census Meeting Minutes. 1963. Meeting minutes dated 20 August 1963. MSS.Archives.2003/1.NHBS 18. Series I. File 140/5, Pt. 1, Demography. First Census N.H. (General), 1962–70, Box C3027052.

NHBS (New Hebrides British Service). 1967a. C.D.1 and C.D. 2. Training Courses, 1–10 March 1967. MSS.Archives.2003/1.NHBS 7. Series IX. File 2, Condominium Census 1967. Instructions to Enumerators in Rural Areas and to Field Supervisors, Box C3025765.

– 1967b. Instructions to Enumerators in Rural Areas Condominium Census, n.d. MSS.Archives.2003/1.NHBS 7. Series IX. File 2, Condominium Census 1967. Instructions to Enumerators in Rural Areas and to Field Supervisors, Box C3025765.

Nicol, James. 1930. Letter to BRC Blandy, 27 April 1930. MSS.Archives.2003/1.NHBS 15. Series I. File 5/1, Pt. 1, Native Court – General, 1916–48, Box C3030550.

– 1934. BDA Report on Native Courts Southern District, 1934. MSS.Archives.2003/1.NHBS 15. Series I. File 5/1, Pt. 1, Native Court – General, 1916–48, Box C3030550.

– 1935. BDA Report on Native Courts Southern District, 1935. MSS. Archives.2003/1.NHBS 15. Series I. File 5/1, Pt. 1, Native Court – General, 1916–48, Box C3030550.

Paton, John. 1941. Letter to BDA G.C. Adams, 29 January 1941. MSS. Archives.2003/1.NHBS 7. Series II. File 3/1, Marriages by Native Customs and Christian Divorce, 1935–50, Box C2453704.

– 1942. Letter to BRC, 12 July 1942. MSS.Archives.2003/1.NHBS 7. Series II. File 3/1, Marriages by Native Customs and Christian Divorce, 1935–50, Box C2453704.

Pyatt, Elizabeth. 1960. *Godden Memorial Hospital Annual Report 1960*. MSS. Archives.2003/1.NHBS 8. Series III. File F8/3/1, Godden Memorial Hospital: General, 1959–66, Box C3026132.

– 1963. *Godden Memorial Hospital Annual Report 1963*. MSS.Archives.2003/ 1.NHBS 8. Series III. File F8/3/1, Godden Memorial Hospital: General, 1959–66, Box C3026132.

– n.d. "Some Vague Ideas on History of Nursing in New Hebrides." Unpublished manuscript in author's possession.

Rees, William. 1966a. Letter to BRC, 19 April 1966; Financial Arrangements at Rural Dresser Stations. MSS.Archives.2003/1.NHBS 18. Series I. File 295/33, Pt. 1, Rural Dresser Service Re-Organisation of, 1965–70, Box C3026287.

– 1966b. Letter to BRC, 19 April 1966; Report: Factors Involved in the Administration of the Rural Health Services at Present Controlled by Paton Memorial Hospital. MSS.Archives.2003/1.NHBS 18. Series I. File 295/33, Pt. 1, Rural Dresser Service Re-Organisation of, 1965–70, Box C3026287.

Townsend, Michael. 1964. Letter to Norma McArthur, 10 October 1964. MSS.Archives.2003/1.NHBS 18. Series I. File 140/3, Pt. 1, Demography. Population Statistics (General), 1961–71, Box C3027052.

Wilkie, Richard. 1965. Letter to Dr. W.J.M Evans, 11 September 1965. MSS. Archives.2003/1.NHBS 18. Series I. File 295/14, Pt. 1, Health Service. P.M.H. Maternity Unit, 1964–6, Box C3027043.

Yaxley, John. 1966. Letter Template for Enumerators from Yaxley, 22 October 1966. MSS.Archives.2003/1.NHBS 18. Series I. File 140/10, Vols. 1–2, Census (N.H. Enumerators and Supervisors, 1966–7), Box C3026840.

– 1967. Letter to Field Supervisors, 22 February 1967. MSS.Archives.2003/ 1.NHBS 18. Series I. File 140/10, Vols. 1–2, Census (N.H. Enumerators and Supervisors, 1966–7), Box C3026840.

Wallmark, S. 1967. Report on Visit to Aoba, Pentecost, and Ambrym, 11–15 April 1967. MSS.Archives.2003/1.NHBS 18. Series I. File 140/10, Vols. 1–2, Census (N.H. Enumerators and Supervisors, 1966–7), Box C3026840.

Index

The letter *f* following a page number denotes a figure; the letter *m*, a map; and the letter *t*, a table.

Anthropological Horizons

Editor: Michael Lambek, University of Toronto

The Varieties of Sensory Experience: A Sourcebook in the Anthropology of the Senses/Edited by David Howes (1991)
Arctic Homeland: Kinship, Community, and Development in Northwest Greenland/ Mark Nuttall (1992)
Knowledge and Practice in Mayotte: Local Discourses of Islam, Sorcery, and Spirit Possession/Michael Lambek (1993)
Deathly Waters and Hungry Mountains: Agrarian Ritual and Class Formation in an Andean Town/Peter Gose (1994)
Paradise: Class, Commuters, and Ethnicity in Rural Ontario/Stanley R. Barrett (1994)
The Cultural World in Beowulf/John M. Hill (1995)
Making It Their Own: Severn Ojibwe Communicative Practices/Lisa Philips Valentine (1995)
Merchants and Shopkeepers: A Historical Anthropology of an Irish Market Town, 1200–1991/P.H. Gulliver and Marilyn Silverman (1995)
Tournaments of Value: Sociability and Hierarchy in a Yemeni Town/Ann Meneley (1996)
Mal'uocchiu: Ambiguity, Evil Eye, and the Language of Distress/Sam Migliore (1997)
Between History and Histories: The Making of Silences and Commemorations/ Edited by Gerald Sider and Gavin Smith (1997)
Eh, Paesan! Being Italian in Toronto/Nicholas DeMaria Harney (1998)
Theorizing the Americanist Tradition / Edited by Lisa Philips Valentine and Regna Darnell (1999)
Colonial 'Reformation' in the Highlands of Central Sulawesi, Indonesia, 1892–1995/ Albert Schrauwers (2000)
The Rock Where We Stand: An Ethnography of Women's Activism in Newfoundland/Glynis George (2000)

'Being Alive Well': Health and the Politics of Cree Well-Being/Naomi Adelson (2000)

Irish Travellers: Racism and the Politics of Culture/Jane Helleiner (2001)

Of Property and Propriety: The Role of Gender and Class in Imperialism and Nationalism/Edited by Himani Bannerji, Shahrzad Mojab, and Judith Whitehead (2001)

An Irish Working Class: Explorations in Political Economy and Hegemony, 1800–1950/Marilyn Silverman (2001)

The Double Twist: From Ethnography to Morphodynamics/Edited by Pierre Maranda (2001)

The House of Difference: Cultural Politics and National Identity in Canada/Eva Mackey (2002)

Writing and Colonialism in Northern Ghana: The Encounter between the LoDagaa and "the World on Paper," 1892–1991/Sean Hawkins (2002)

Guardians of the Transcendent: An Ethnography of a Jain Ascetic Community/Anne Vallely (2002)

The Hot and the Cold: Ills of Humans and Maize in Native Mexico/Jacques M. Chevalier and Andrés Sánchez Bain (2003)

Figured Worlds: Ontological Obstacles in Intercultural Relations/Edited by John Clammer, Sylvie Poirier, and Eric Schwimmer (2004)

Revenge of the Windigo: The Construction of the Mind and Mental Health of North American Aboriginal Peoples/James B. Waldram (2004)

The Cultural Politics of Markets: Economic Liberalization and Social Change in Nepal/Katharine Neilson Rankin (2004)

A World of Relationships: Itineraries, Dreams, and Events in the Australian Western Desert/Sylvie Poirier (2005)

The Politics of the Past in an Argentine Working-Class Neighbourhood/Lindsay DuBois (2005)

Youth and Identity Politics in South Africa, 1990–1994/Sibusisiwe Nombuso Dlamini (2005)

Maps of Experience: The Anchoring of Land to Story in Secwepemc Discourse/Andie Diane Palmer (2005)

We Are Now a Nation: Croats between 'Home' and 'Homeland'/Daphne N. Winland (2007)

Beyond Bodies: Rain-Making and Sense-Making in Tanzania/Todd Sanders (2008)

Kaleidoscopic Odessa: History and Place in Contemporary Ukraine/Tanya Richardson (2008)

Invaders as Ancestors: On the Intercultural Making and Unmaking of Spanish Colonialism in the Andes/Peter Gose (2008)

From Equality to Inequality: Social Change among Newly Sedentary Lanoh Hunter-Gatherer Traders of Peninsular Malaysia/Csilla Dallos (2011)

Rural Nostalgias and Transnational Dreams: Identity and Modernity among Jat Sikhs/Nicola Mooney (2011)
Dimensions of Development: History, Community, and Change in Allpachico, Peru/Susan Vincent (2012)
People of Substance: An Ethnography of Morality in the Colombian Amazon/Carlos David Londoño Sulkin (2012)
'We Are Still Didene': Stories of Hunting and History from Northern British Columbia/Thomas McIlwraith (2012)
Being Māori in the City: Indigenous Everyday Life in Auckland/Natacha Gagné (2013)
The Hakkas of Sarawak: Sacrificial Gifts in Cold War Era Malaysia/Kee Howe Yong (2013)
Remembering Nayeche and the Gray Bull Engiro: African Storytellers of the Karamoja Plateau and the Plains of Turkana/Mustafa Kemal Mirzeler (2014)
In Light of Africa: Globalizing Blackness in Northeast Brazil/Allan Charles Dawson (2014)
The Land of Weddings and Rain: Nation and Modernity in Post-Socialist Lithuania/Gediminas Lankauskas (2015)
Milanese Encounters: Public Space and Vision in Contemporary Urban Italy/Cristina Moretti (2015)
Legacies of Violence: History, Society, and the State in Sardinia/Antonio Sorge (2015)
Looking Back, Moving Forward: Transformation and Ethical Practice in the Ghanaian Church of Pentecost/Girish Daswani (2015)
Why the Porcupine Is Not a Bird: Explorations in the Folk Zoology of an Eastern Indonesian People/Gregory Forth (2016)
The Heart of Helambu: Ethnography and Entanglement in Nepal/Tom O'Neill (2016)
Tournaments of Value: Sociability and Hierarchy in a Yemeni Town, 20th Anniversary Edition/Ann Meneley (2016)
Europe Un-Imagined: Nation and Culture at a French-German Television Channel/Damien Stankiewicz (2017)
Transforming Indigeneity: Urbanization and Language Revitalization in the Brazilian Amazon/Sarah Shulist (2018)
Wrapping Authority: Women Islamic Leaders in a Sufi Movement in Dakar, Senegal/Joseph Hill (2018)
Island in the Stream: An Ethnographic History of Mayotte/Michael Lambek (2018)
Materializing Difference: Consumer Culture, Politics, and Ethnicity among Romanian Roma/Péter Berta (2019)
Virtual Activism: Sexuality, the Internet, and a Social Movement in Singapore/Robert Phillips (2020)
Shadow Play: Information Politics in Urban Indonesia/Sheri Lynn Gibbins (2021)

Suspect Others: Spirit Mediums, Self-Knowledge, and Race in Multiethnic Suriname / Stuart Earle Strange (2021)

Exemplary Life: Modeling Sainthood in Christian Syria / Andreas Bandak (2022)

Without the State: Self-Organization and Political Activism in Ukraine/Emily Channell-Justice (2022)

Moral Figures: Making Reproduction Public in Vanuatu / Alexandra Widmer (2023)

Truly Human: Indigeneity and Indigenous Resurgence on Formosa / Scott Simon (2023)

Milton Keynes UK
Ingram Content Group UK Ltd.
UKHW012143210424
441411UK00002B/24